SpringerBriefs in Applied Sciences and Technology

Manufacturing and Surface Engineering

Series Editor

Joao Paulo Davim, Department of Mechanical Engineering, University of Aveiro, Aveiro, Portugal

This series fosters information exchange and discussion on all aspects of manufacturing and surface engineering for modern industry. This series focuses on manufacturing with emphasis in machining and forming technologies, including traditional machining (turning, milling, drilling, etc.), non-traditional machining (EDM, USM, LAM, etc.), abrasive machining, hard part machining, high speed machining, high efficiency machining, micromachining, internet-based machining, metal casting, joining, powder metallurgy, extrusion, forging, rolling, drawing, sheet metal forming, microforming, hydroforming, thermoforming, incremental forming, plastics/composites processing, ceramic processing, hybrid processes (thermal, plasma, chemical and electrical energy assisted methods), etc. The manufacturability of all materials will be considered, including metals, polymers, ceramics, composites, biomaterials, nanomaterials, etc. The series covers the full range of surface engineering aspects such as surface metrology, surface integrity, contact mechanics, friction and wear, lubrication and lubricants, coatings an surface treatments, multiscale tribology including biomedical systems and manufacturing processes. Moreover, the series covers the computational methods and optimization techniques applied in manufacturing and surface engineering. Contributions to this book series are welcome on all subjects of manufacturing and surface engineering. Especially welcome are books that pioneer new research directions, raise new questions and new possibilities, or examine old problems from a new angle. To submit a proposal or request further information, please contact Dr. Mayra Castro, Publishing Editor Applied Sciences, via mayra.castro@springer.com or Professor J. Paulo Davim, Book Series Editor, via pdavim@ua.pt

More information about this subseries at http://www.springer.com/series/10623

Pawel Pawlus · Andrzej Dzierwa ·
Agnieszka Lenart

Dry Gross Fretting of Rough Surfaces

Influential Parameters

 Springer

Pawel Pawlus 🆔
Department of Manufacturing Processes
and Production Engineering
Rzeszów University of Technology
Rzeszów, Poland

Andrzej Dzierwa 🆔
Department of Manufacturing Processes
and Production Engineering
Rzeszów University of Technology
Rzeszów, Poland

Agnieszka Lenart
EME Aero Sp. z o. o.
Jasionka, Poland

ISSN 2191-530X   ISSN 2191-5318   (electronic)
SpringerBriefs in Applied Sciences and Technology
ISSN 2365-8223   ISSN 2365-8231   (electronic)
Manufacturing and Surface Engineering
ISBN 978-3-030-31562-7   ISBN 978-3-030-31563-4   (eBook)
https://doi.org/10.1007/978-3-030-31563-4

This Springer imprint is published by the registered company Springer Nature Switzerland AG
The registered company address is: Gewerbestrasse 11, 6330 Cham, Switzerland

Introduction

It is well known that surface topography affects functional properties of machine parts, such as material contact, sealing, friction, lubrication and wear resistance. Therefore, many of the demands in industry are concerned with the surface of a component. The surface topography is generated in the final stages of machining processes. One can say that the surface topography is the fingerprint of the manufacturing. Although there are a lot of scientific works within this area, the knowledge about the relationship between surface topography and surface function is still incomplete. Most of researchers analysed the relations between surface height and functional properties of machine components. However, the other surface features like spatial and hybrid also affect functions of machine parts. There are strong connections between surface topography and tribological properties of sliding parts, especially in the initial stage of machine life.

A reciprocal relative motion between two surfaces in contact is often observed. If the relative displacement is smaller than the contact size, fretting occurs. It can be defined as a relative motion with a very small amplitude between two oscillating surfaces. Depending on the operating conditions, various sliding regimes can be identified: partial slip and gross slip. For a very small displacement, partial slip occurs, which is primarily connected with cracking observed near contact edges. In this case, fretting fatigue takes places. When the relative displacements are so large to cause a gross slip between surfaces in contact, the material removal, fretting wear occurs. Because the relative displacement is lower from size of the contact, the role of wear debris is substantial. Worn particles can be entrapped within the contact zone or ejected. Fretting combines various wear mechanisms, including mainly abrasion and adhesion. Fretting was observed in various industries. Therefore, an attempt to reduce fretting wear and the resistance to motion is a problem of a great functional importance.

There are many operating parameters affecting fretting wear. From among of them the typical are: amplitude, normal load and frequency. The effect of frequency on fretting wear is still disputable. In only a few works, the effect of frequency on the friction force was investigated.

The surface machining is essential to reduce damage caused by fretting; however, typically effects of surface topography are mostly neglected. There is an opinion that on a rough surface abrasive particles can escape from contact zone to neighbouring valleys, instead of ploughing; therefore, very rough surface is needed. However, the results of the effect of surface topography on wear are often contradictory; in most of research, only the effect of surface amplitude on friction and wear in fretting was studied.

This book is the attempt to rectify the presented above problems. The effect of surface topography of steel disc and operating parameters on friction and wear under gross slip conditions in fretting contact between steel disc and ball was experimentally studied.

Pawel Pawlus
Andrzej Dzierwa
Agnieszka Lenart

Contents

1 **Surface Topography and Its Functional Importance** 1
   1.1   Surface Topography Analysis 1
   1.2   Functional Importance of Surface Topography 4
   References 13

2 **Fretting** 17
   2.1   Fundamental Information 17
   2.2   The Effects of Operating Parameters on Fretting 21
   2.3   The Effect of Surface Texture on Fretting 22
   References 23

3 **The Effect of Frequency and Normal Load on Dry Gross**
   Fretting of Rough Surfaces 27
   3.1   Experimental Conditions 27
   3.2   Results and Discussion 29
   References 54

Concluding Remarks 57

Chapter 1
Surface Topography and Its Functional Importance

1.1 Surface Topography Analysis

The 2D profile was analysed earlier. Standards ISO 4287 and ISO 11562 define the roughness profile, waviness profile, the primary profile and their parameters. The mean line is determined by fitting a least squares line through the profile. The roughness profile is derived from the primary profile by retaining the short-wave component. The waviness profile contains long wavelengths. Figure 1.1 shows schematically the error of shape, primary profile as well as roughness and waviness profiles. The parameters calculated from the primary profile are called P-parameters, from the roughness profile—R-parameters, from the waviness profile—W–parameters. The parameters can be divided into: height, spacing, hybrid and other parameters (e.g. derived from material ratio curve).

In the early 1980s, academic community started with the surface characterisation in 3 dimensions. Now, the 3D (areal) surface topography analysis is the common practice. Surface topography comprises error of shape, waviness and roughness. Surface texture means waviness and roughness. It is obtained after form removal using typically polynomials (Fig. 1.2).

Traditional texture parameters use a statistical basis to characterise 3D surface topography. These are termed "field parameters". They consist of the S-parameter set and the V-parameter set. The S-parameters are based on the height and spatial frequency. The V-parameters give fundamental information of a material ratio curve. The S-parameter set contains 15 parameters and was classified into five types: amplitude, spacing, hybrid and other parameters.

Root-mean-square deviation of the surface **Sq** is defined as a root-mean-square value of the surface departures within the sampling area.

Skewness of topography distribution **Ssk** measures asymmetry of surface deviations about the mean/reference plane.

© The Author(s), under exclusive license to Springer Nature Switzerland AG 2020
P. Pawlus et al., *Dry Gross Fretting of Rough Surfaces*, Manufacturing
and Surface Engineering, https://doi.org/10.1007/978-3-030-31563-4_1

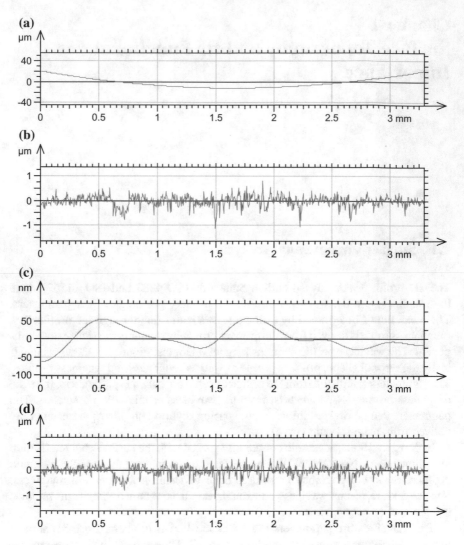

Fig. 1.1 Components of the surface topography: error of shape (**a**), primary profile (**b**), waviness profile (**c**) and roughness profile (**d**)

Kurtosis of the topography distribution **Sku** is a measure of the sharpness of the surface height distribution. The Ssk parameter represents the symmetry of the surface, while the Sku parameter spreading distribution.

The maximum surface peak height **Sp** is the largest peak height value from the reference surface within the sampling area.

The lowest valley of the surface **Sv** is the largest valley depth value from the reference surface within the sampling area.

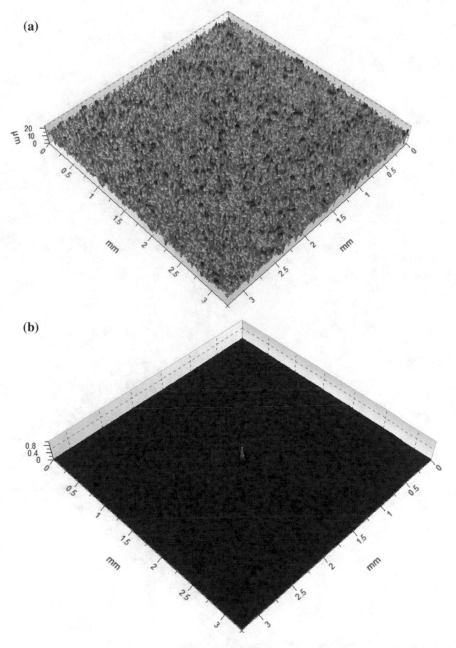

Fig. 1.3 Isotropic surface and its autocorrelation function

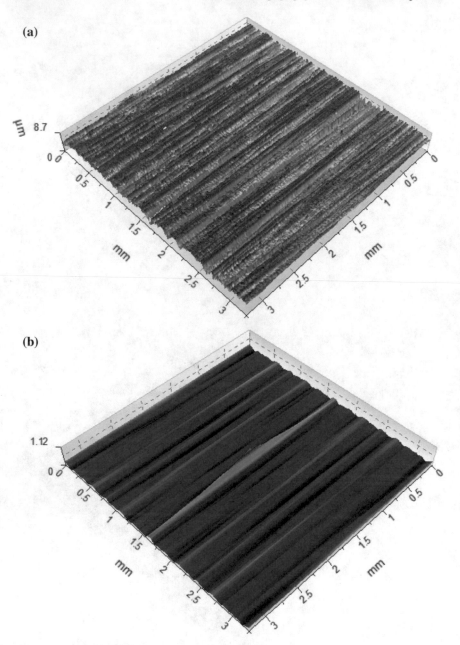

Fig. 1.4 Anisotropic surface and its autocorrelation function

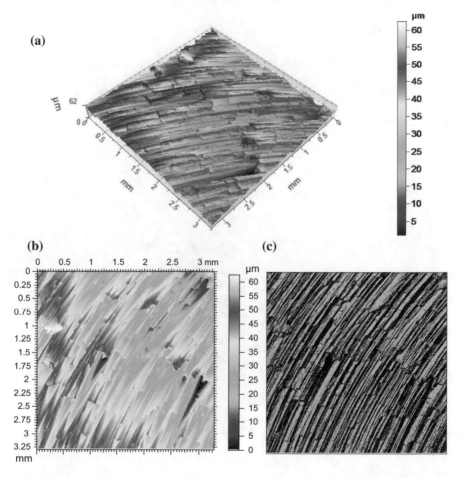

Fig. 1.5 Various methods of surface topography presentation: **a** isometric view, **b** contour plot and **c** surface photograph

plasticity index, which depends on both material and topographic parameters. The well-known version of plasticity index was defined by Greenwood and Williamson (GW) [15]:

$$\Psi = \frac{E'}{H}\left(\frac{\sigma_s}{R}\right)^{1/2}$$

where H—hardness of softer material, E'—Young's elastic modulus, σ_s—standard deviation of asperity heights and R—mean radius of summit curvatures.

From among surface texture parameters, the standard deviation of asperity heights and the average radius of asperity curvatures were considered. It is necessary to distinguish between peak heights, summit heights and surface ordinate heights. Summits

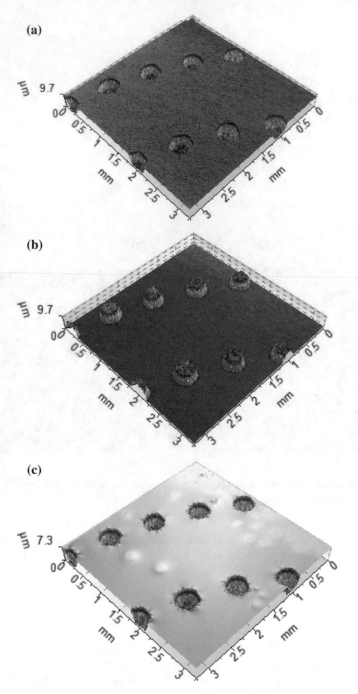

Fig. 1.6 Isometric views of measured textured surface (**a**), surface inverted (**b**) and surface truncated (**c**)

Fig. 1.7 Connection
between surface topography,
manufacturing and function

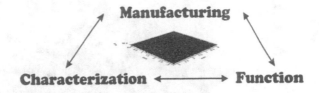

are local maxima on a surface, while peaks are local maxima on a profile from surface
[49]. According to [15], the contact mode was elastic for plasticity index lower than
0.6, plastic for higher than 1.0, while for the range 0.6–1, the mode of deformation
was not clear. This index was extended for anisotropic surfaces [5, 32]. Whitehouse
and Archard (WA) [50] developed their version of plasticity index taking into consideration 2D profile properties like a standard deviation of surface height and a
correlation length (the distance at which the autocorrelation function decayed to a
given value). In technical literature, there are also plasticity index versions based on
rms. slope [13, 46], but Greenwood and Williamson approach is the most commonly
used. Hirst and Hollander [16] as well as Poon and Sayles [42] found that boundary
from unsafe to safe sliding depended on the plasticity index.

Mechanics of contact between bodies depends on surface topography of components. Greenwood and Williamson [15] firstly developed an elastic contact model
(GW model) of an isotropic Gaussian surface—it was extended to cover two rough
surfaces [14]. Abbott and Firestone [1] on the other side established a model for fully
plastic contact. Pullen and Williamson [43] for plastic contact developed a volume
conservation model. This model was the base of an elastic–plastic contact model
(CEB) [4]. Zhao and Maietta developed a different elastic–plastic contact model
(ZMC) for contact of rough surfaces [55]. Many publications were inspired by CEB
and ZMC models. Novel models are based on FEM analysis. Kogut and Etsion [22,
23] performed an FEM analysis of contact between an elastic–plastic and a rigid
flat (KE model) in full slip condition. In Jackson and Green works [19, 20], a FEM
analysis was performed which accounted for geometry and material effects that were
not included in KE approach. The other CKE model [6] assumed full-stick condition
at the asperity level.

The presented above models of rough surfaces contact are based on a simple statistical assumption of Gaussian distribution of asperity heights. However, surfaces
after a low wear (within the limit of the original surface topography) seldom display
such feature. Even for initial one-process symmetric roughness, the running-in process often resulted in asymmetric height distribution. Two-process stratified surface
is created. This kind of surface was formed by plateau honing process. This type of
surface resembles surface created during a low wear process. It consists of a smooth
plateau and wide and deep valleys. It combines good frictional properties of a smooth
surface with a great ability to maintain lubricant by a porous structure—Fig. 1.8.

Two-process surfaces have advantages over conventional one-process textures
[3, 8, 12, 21]—smaller friction and higher wear resistance. There were only a few
publications concerning the study of contact of two-process surfaces [7, 29, 39, 40,
47]. Leefe predicted elastic contact of two-process surfaces [29]. Tomanik computed

(a)

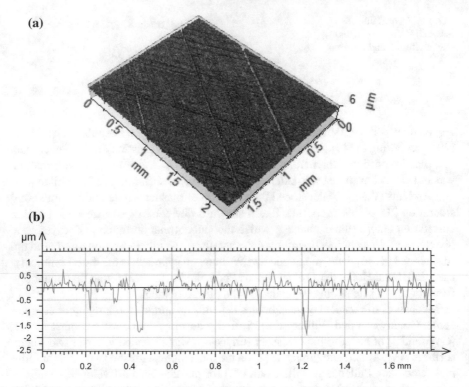

(b)

Fig. 1.8 Isometric view (**a**) and extracted profile (**b**) of plateau honed cylinder surface

elastic contact parameters for highly skewed cylinder surface [47]. Dimkovski used elastic and elastic–plastic models for measured plateau honed cylinder liner surfaces [7]. Pawlus et al. [39, 40] developed plasticity indices for two-process textures.

Generally, during friction in the presence of lubrication, the smooth surface has small ability to maintain oil; however, wear of rough texture is large (see Fig. 1.9). Therefore, the machined surface should be not too smooth and not too rough—the selection of the surface amplitude should be a compromise. Machined surface should be similar to the surface created during running-in (Fig. 1.10). In this case, duration of

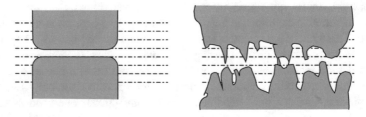

Fig. 1.9 Effect of too smooth and too rough surface on operating conditions in the presence of lubrication

Fig. 1.10 Change of texture height during running-in

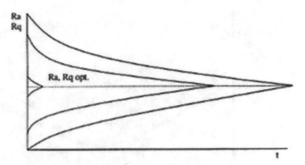

the running-in is shorter and wear during running-in is smaller. The effect of surface topography is especially important during running-in or a low wear. However, this influence on stabilised wear can be also important [38].

The effect of surface topography on tribological performance in mixed and boundary lubrication is larger than in full fluid film lubrication. The texture height affects the type of lubrication. When the oil film thickness is larger than roughness height, the influence of surface texture on lubrication is not substantial. During fluid lubrication, smooth texture is preferred. However, in this case, the high peaks may cause contact between surfaces, as the effect of it the oil film can be broken.

The wear intensity is often proportional to initial surface height [28, 31, 44]. During hydrodynamic lubrication, the position of the surface to the movement direction is substantial. Patir and Cheng [37] analysed theoretically the effect of surface anisotropy on oil flow in partial HD conditions. The surfaces transversely oriented did not cause flow resistance, and the side flow was low. For isotropic surfaces, main and side flows had similar values. For surfaces longitudinally oriented, both side flow and main flow increased.

The effects of roughness height on the resistance on static and kinetic friction are opposite. In the static friction, the coefficient of friction increased when the roughness height is getting smaller [24, 51]. This dependence was caused by an increase in a real area of contact.

During dry friction between metals during reduction of the roughness height, the coefficient of friction initially decreased, obtained minimum value and then increased (see Fig. 1.11). This dependence is caused by a decrease in mechanical (μ_m) and an increase on molecular (adhesive) (μ_a) frictional resistance during a decrease in the roughness height.

When lubricant is added, typically the coefficient of friction is larger when roughness height is bigger [30, 36].

They are some approaches to decrease in the resistance to motion by changing surface texture of machine elements. The introduction of specific textures on sliding surfaces, including micropits (or holes, dimples, cavities, oil pockets), is one of these approaches [10, 35]. Plateau honing is the earliest example of surface texturing. Textured surfaces led to improvement of the seizure resistance of sliding elements [11, 25]. Holes can also serve as microtraps for wear debris in lubricated or dry

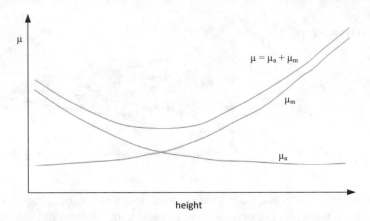

Fig. 1.11 Effect of roughness height in the coefficient of friction

sliding [10, 35]. Various techniques can be employed for surface texturing including machining, ion beam texturing, etching techniques and laser texturing, which is the most popular technique [9, 10]. Burnishing [11, 25] and abrasive jet machining [34, 48, 52] are promising methods.

Surface texturing has been shown to provide tribological benefits in terms of friction reduction in conformal contact [18, 27, 33, 52]. It is possible to select parameters of textured surfaces like the oil pockets density, sizes of dimples and surface roughness in areas free of dimples based on the literature analysis. However, the effect of oil pockets array is also of a great tribological importance [54]. Optimisation of dimple pattern is an approach to enhance the tribological effect on the whole sliding surface, since two adjacent dimples could have interactions with each other. The spiral oil pocket array is recommended on the basis of experimental research [18]. Dimples are typically of spherical shape. However, numerical investigation revealed that dimples of rectangular and triangular shapes substantially affected the friction force [2]. Chevron like oil pockets are also promising [53]. Figure 1.12 shows textures of various shapes of oil pockets.

Typically, the effect of surface texturing is larger in regime of mixed compared to fluid lubrication. However, a textured surface can be designed for both full and mixed lubrications with relatively large and shallow oil pockets for full lubrication and smaller dimples of higher depth for mixed lubrication [17].

Surface texturing may be combined with coating. A surface coated with TiN and DLC layers [41] before texturing led to prolonged lifetime and reduced friction. A combination of DLC coating and texturing of cylinder liner surface caused improvement of effective power in internal combustion engine [26].

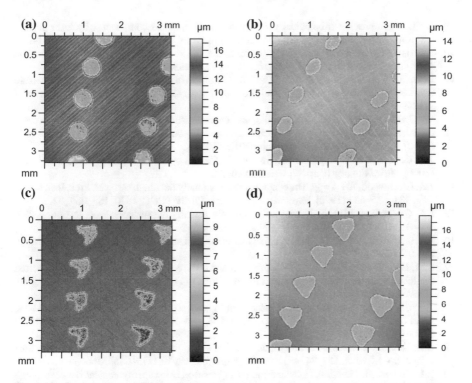

Fig. 1.12 Contour plots of different oil pockets shapes, **a** circle; **b** oval; **c** chevron; **d** triangle

References

1. E.J. Abbott, F.A. Firestone, Specifying surface quality—a method based on accurate measurement and comparison. Mech. Eng. **55**, 569–572 (1933)
2. L. Burstein, D. Ingman, Pore ensemble statistics in application to lubrication under reciprocating motion. Tribol. Trans. **43**(2), 205–212 (2000). https://doi.org/10.1080/10402000008982330
3. J.C. Campbell, Cylinder bore surface roughness in internal combustion engines: its appreciation and control. Wear **19**, 163–166 (1972). https://doi.org/10.1016/0043-1648(72)90302-X
4. W.R. Chang, I. Etsion, D.B. Bogy, An elastic-plastic model for the contact of rough surfaces. J. Tribol. Trans. ASME **109**, 257–263 (1987). https://doi.org/10.1115/1.3261348
5. W. Chengwei, Z. Linqing, A general expression for plasticity index. Wear **121**, 161–172 (1988). https://doi.org/10.1016/0043-1648(88),90040-3
6. O. Cohen, Y. Kligerman, I. Etsion, A model for contact and static friction of nominally flat rough surfaces under full stick contact condition. J. Tribol. Trans. ASME **130**(3), 031401 (2008). https://doi.org/10.1115/1.2908925
7. Dimkovski Z (2011) Surfaces of honed cylinder liners. PhD Dissertation, Chalmers University of Technology
8. A. Dzierwa, P. Pawlus, W. Zelasko et al., The study of the tribological properties of one-process and two-process surfaces after vapour blasting and lapping using pin-on-disc tester. Key Eng. Mater. **527**, 217–222 (2013). https://doi.org/10.4028/www.scientific.net/KEM.527.217

9. I. Etsion, Improving tribological performance of mechanical components by laser surface texturing. Tribol. Lett. **17**(4), 733–737 (2004). https://doi.org/10.1007/s11249-004-8081-1

10. I. Etsion, State of the art in laser surface texturing, in *Proceedings of the 12th Conference on Metrology and Properties of Engineering Surfaces*, Rzeszow, Poland, 08–10 July 2009

11. L. Galda, A. Dzierwa, J. Sep et al., The effect of oil pockets shape and distribution on seizure resistance in lubricated sliding. Tribol. Lett. **37**(2), 301–311 (2010). https://doi.org/10.1007/s11249-009-9522-7

12. W. Grabon, P. Pawlus, J. Sep, Tribological characteristics of one-process and two-process cylinder liner honed surfaces under reciprocating sliding conditions. Tribol. Int. **43**, 1882–1892 (2010). https://doi.org/10.1016/j.triboint.2010.02.003

13. J.A. Greenwood, A simplified elliptic model of rough surface contact. Wear **261**, 191–200 (2006). https://doi.org/10.1016/j.wear.2005.09.031

14. J.A. Greenwood, J.H. Tripp, The contact of two nominally flat rough surfaces. Proc. Inst. Mech. Eng. **185**, 625–633 (1970). https://doi.org/10.1243/PIME_PROC_1970_185_069_02

15. J.A. Greenwood, J.B.P. Williamson, Contact of nominally flat surfaces. Proc. R. Soc. A **295**, 300–319 (1966). https://doi.org/10.1098/rspa.1966.0242

16. W. Hirst, A.E. Hollander, Surface finish and damage in sliding. Proc. R. Soc. Lond. A Math. Phys. Sci. **337A**, 379–394 (1974)

17. S.M. Hsu, Y. Jing, F. Zhao, Self-adaptive surface texture design for friction reduction across the lubrication regimes. Surf. Topogr. Metrol. Prop. **4**(1), 014004 (2015). https://doi.org/10.1088/2051-672x/4/1/014004

18. T. Hu, L. Hu, Q. Ding, The effect of laser texturing on the tribological behaviour of Ti-6Al-4V. Proc. Inst. Mech. Eng. Part J J. Eng. Tribol. **226**, 854–863 (2012). https://doi.org/10.1177/1350650112450801

19. R.L. Jackson, I. Green, A finite element study of elasto-plastic hemispherical contact against a rigid flat. J. Tribol. **127**, 343–354 (2005). https://doi.org/10.1115/1.1866166

20. R.L. Jackson, I. Green, A statistical model of elasto-plastic asperity contact between rough surfaces. Tribol. Int. **39**, 906–914 (2006). https://doi.org/10.1016/j.triboint.2005.09.001

21. Y. Jeng, Impact of plateaued surfaces on tribological performance. Tribol. Trans. **39**, 354–361 (1996). https://doi.org/10.1080/10402009608983538

22. L. Kogut, I. Etsion, Elastic-plastic contact analysis of a sphere and a rigid flat. J. Appl. Mech. **69**, 657–662 (2002). https://doi.org/10.1115/1.1490373

23. L. Kogut, T. Etsion, A finite element based elastic-plastic model for the contact of rough surfaces. Tribol. Trans. **46**, 383–390 (2003). https://doi.org/10.1080/10402000308982641

24. R. Koka, T. Pitchford, M. Jesh et al., Studies on head-disc contact increase during contact start/stop and continuous drag testing of thin film disc. Tribol. Trans. **36**(1), 1–10 (1993). https://doi.org/10.1080/10402009308983125

25. W. Koszela, L. Galda, A. Dzierwa et al., The effect of surface texturing on seizure resistance of a steel-bronze assembly. Tribol. Int. **43**, 1933–1942 (2010). https://doi.org/10.1016/j.triboint.2010.04.016

26. W. Koszela, P. Pawlus, R. Reizer et al., The combined effect of surface texturing and DLC coating on the functional properties of internal combustion engines. Tribol. Int. **127**, 470–477 (2018). https://doi.org/10.1016/j.triboint.2018.06.034

27. O. Kovalchenko, A. Ajayi, A. Erdemir et al., The effect of laser surface texturing on transitions in lubrication regimes during unidirectional sliding contact. Tribol. Int. **38**, 219–225 (2005). https://doi.org/10.1016/j.triboint.2004.08.004

28. Z. Krzyzak, P. Pawlus, 'Zero-wear' of piston skirt surface topography. Wear **260**, 554–561 (2006). https://doi.org/10.1016/j.wear.2005.03.038

29. S.E. Leefe, "Bi-Gaussian" representation of worn surface topography in elastic contact problems, in *Tribology for Energy Conservation*, ed. by D. Dowson, et al. (Imperial College of Science, Technology and Medicine, London, 1998), pp. 281–290

30. Z. Ma, N.E. Henein, W. Bryzik et al., Break-in liner wear and piston ring assembly friction in a spark–ignited engine. Tribol. Trans. **41**, 497–504 (1998). https://doi.org/10.1080/10402009808983774
31. G. Masouros, A. Dimarogonas, K. Lefas, A model for wear and surface roughness transients during the running-in of bearings. Wear **45**, 375–383 (1977). https://doi.org/10.1016/0043-1648(77)90028-X
32. B.B. Mihic, Thermal contact conductance: theoretical considerations. Int. J. Heat Mass Transf. **17**, 205–214 (1974). https://doi.org/10.1016/0017-9310(74),90082-9
33. S.P. Mishra, A.A. Polycarpou, Tribological studies of unpolished laser surface textures under starved lubrication conditions for use in air-conditioning and refrigeration compressors. Tribol. Int. **44**(12), 1890–1901 (2011). https://doi.org/10.1016/j.triboint.2011.08.005
34. M. Nakano, M. Korenaga, K. Miyake et al., Applying micro-texture to cast iron surfaces to reduce the friction coefficient under lubricated conditions. Tribol. Lett. **28**, 131–138 (2007). https://doi.org/10.1007/s11249-007-9257-2
35. B. Nilsson, B.G. Rosen, T.R. Thomas et al., Oil pockets and surface topography: mechanism of friction reduction, in *Abstracts of the XI International Colloquium on Surfaces*, Chemnitz Germany, 2–3 Feb 2004
36. I. Nogueira, A.M. Dias, R. Gras et al., An experimental model for mixed friction during running-in. Wear **253**, 541–549 (2002). https://doi.org/10.1016/S0043-1648(02),00065-0
37. N. Patir, H.S. Cheng, An average flow model for determining effects of three dimensional roughness on partial hydrodynamic lubrication. J. Lubr. Technol. **100**(1), 12–17 (1978). https://doi.org/10.1115/1.3453103
38. P. Pawlus, Effects of honed cylinder surface topography on the wear of piston-piston ring-cylinder assemblies under artificially increased dustiness conditions. Tribol. Int. **26**(1), 49–55 (1993). https://doi.org/10.1016/0301-679X(93),90038-3
39. P. Pawlus, W.A. Grabon, D. Czach, Calculation of plasticity index of honed cylinder liner textures. J. Phys. Conf. Ser. **1183**(1), 012003 (2019). https://doi.org/10.1088/1742-6596/1183/1/012003
40. P. Pawlus, W. Zelasko, R. Reizer et al., Calculation of plasticity index of two-process surfaces. Proc. Inst. Mech. Eng. Part J J. Eng. Tribol. **231**, 572–582 (2017). https://doi.org/10.1177/1350650116664826
41. U. Pettersson, S. Jacobson, Influence of surface texture on boundary lubricated sliding contacts. Tribol. Int. **36**(11), 857–864 (2003). https://doi.org/10.1016/S0301-679X(03)00104-X
42. C.Y. Poon, R.S. Sayles, The classification of rough surface contacts in relation to tribology. J. Phys. D Appl. Phys. **25**(1A), 249–256 (1992). https://doi.org/10.1088/0022-3727/25/1A/038
43. J. Pullen, J.B.P. Williamson, On the plastic contact of rough surfaces. Proc. R. Soc. Lond. A Math. Phys. Sci. **A327**, 159–173 (1972). https://doi.org/10.1098/rspa.1972.0038
44. B.J. Roylance, C.H. Bovington, G. Wang, A. Hubbard, Running-in wear behaviours of valve train system, in *Vehicle Tribology*, ed. by D. Dowson et al. Tribology Series, vol. 18 (1991), pp. 143–147
45. T.R. Thomas, *Rough Surfaces* (Imperial College Press, London, 1999)
46. T.R. Thomas, B.G. Rosen, Determination of the optimum sampling interval for rough contact mechanics. Tribol. Int. **33**, 601–610 (2000). https://doi.org/10.1016/S0301-679X(00),00076-1
47. E. Tomanik, Modelling the asperity contact area on actual 3D surfaces. SAE Technical Paper 2005-01-1864 (2005). https://doi.org/10.4271/2005-01-1864
48. M. Wakuda, Y. Yamauchi, S. Kanzaki et al., Effect of surface texturing on friction reduction between ceramic and steel materials under lubricated sliding contact. Wear **254**(3–4), 356–363 (2003). https://doi.org/10.1016/S0043-1648(03),00004-8
49. D.J. Whitehouse, *Handbook of surface Metrology* (Institute of Physics Publishing, Bristol and Philadelphia, 1994)
50. D.J. Whitehouse, J.F. Archard, The properties of random surface of significance in their contact. Proc. R. Soc. Lond. A Math. Phys. Sci. **A316**, 97–121 (1970)

51. K.L. Woo, T.R. Thomas, Roughness, friction and wear: the effect of contact planform. Wear **57**, 357–363 (1979). https://doi.org/10.1016/0043-1648(79),90109-1

52. S. Wos, W. Koszela, P. Pawlus, Determination of oil demand for textured surfaces under conformal contact conditions. Tribol. Int. **93**, 602–613 (2016). https://doi.org/10.1016/j.triboint.2015.05.016

53. S. Wos, W. Koszela, P. Pawlus, The effect of the shape of oil pockets on the friction force. Tribologia **4**, 151–156 (2018). https://doi.org/10.5604/01.3001.0012.7556

54. H. Yu, W. Huang, X. Wang, Dimple patterns design for different circumstances. Lubr. Sci. **25**, 67–78 (2011). https://doi.org/10.1002/ls.168

55. Y. Zhao, D.M. Maietta, L. Chang, An asperity microcontact model incorporating the transition from elastic deformation to fully plastic flow. J. Tribol. **20**, 86–93 (2000). https://doi.org/10.1115/1.555332

Chapter 2
Fretting

2.1 Fundamental Information

Fretting can be defined as a relative motion with a very small amplitude between two oscillating surfaces. Depending on the applied normal load and relative displacements, various sliding regimes can be identified: a partial slip and a gross slip [49, 56]. A fretting damage type is dependent on sliding regimes, and it leads to cracking (fretting fatigue) under a partial slip (lower amplitude) and wear damage under a gross slip (higher amplitude)—see Fig. 2.1. When the relative displacement is larger than the contact size, sliding wear occurs. Various modes of fretting can be identified on the basis of the shape of fretting loop. When it is similar to rhombus, gross slip appears, while it is similar to ellipse, partial slip takes place (Fig. 2.2). In order to distinguish between various fretting modes and reciprocating sliding various parameters like a slip index, an energy ratio and a slip ratio can be used (Fig. 2.3).

One can calculate the slip index according to the following equation [46, 47]:

$$\delta = \frac{\tan S_c A_d}{P} = \frac{Q A_d}{P(A_t - A_d)}$$

where

A_d—actual amplitude of displacement,
A_t—total amplitude of displacement,
S_c—slope of the fretting loop,
Q—tangential force,
P—normal force.

It was found that the gross slip regime was characterised by $0.6 < \delta < 10$, the partial slip regime by $0.5 < \delta < 0.6$, while for $\delta > 11$ reciprocating sliding occurred [46, 47]. The slip ratio A_t/A_d is also useful for identifying a fretting regime. A transition from the partial to the gross slip was 0.26 or 0.17, but 0.95 between the gross slip and reciprocating sliding [5, 46]. The energy ratio is the ratio of dissipated energy

P. Pawlus et al., *Dry Gross Fretting of Rough Surfaces*, Manufacturing and Surface Engineering, https://doi.org/10.1007/978-3-030-31563-4_2

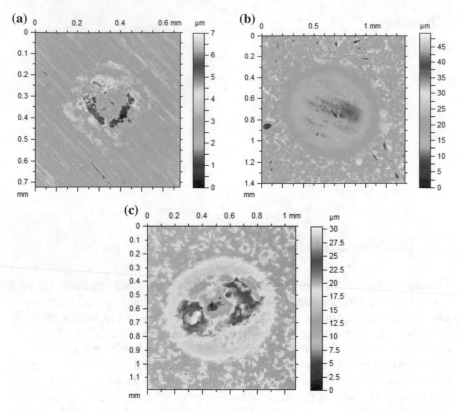

Fig. 2.1 Various types of fretting: partial slip (**a**), gross slip (**b**, **c**), when dominant wear mechanism is abrasion (**b**) and adhesion (**c**)

(within fretting loop E_d) to the total energy E_t (Fig. 2.3). When the ratio E_d/E_t is larger than 0.2, the test is completed within the gross slip regime [5]. When the relative displacement is higher than the contact size, sliding wear occurs. Zhou and Vincent [57] established also mixed fretting regime corresponding to the transition between the gross slip to the partial slip.

Fretting wear combines various basic types of wear, including mainly adhesion and abrasion (Fig. 2.1) as well as surface fatigue and oxidation. In fretting wear, the generated oxide debris is of huge importance—this is a characteristic feature of fretting. Two different effects of wear debris on material loss were typically identified—harmful due to the abrasive action or beneficial thanks to formation of the oxide layer; they are dependent on the kind of wear (adhesion or abrasion) [11]. During fretting wear, particles are trapped in the fretted regions and undergo a series of processes before they are removed. Lemm et al. found that when hardness between two steel specimens was high, the oxide-based particles were retained on the softer material; this resulted in protection of the softer specimen and substantial wear removal of the harder counterspecimen [23].

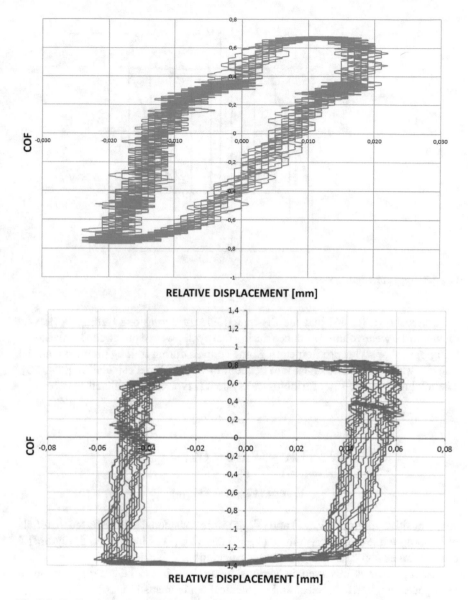

Fig. 2.2 Fretting loops for partial slip (upper graph) and gross slip (lower graph)

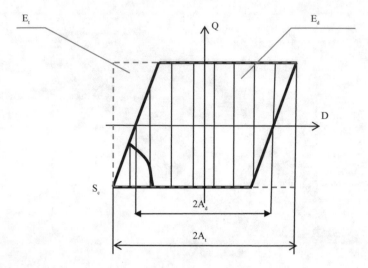

Fig. 2.3 Typical fretting loop

It was shown [6, 54] that the dissipated energy during one fretting cycle was related to the wear volume produced during that cycle. Wear damage can be measured using optical or stylus profilometers. Some couples during fretting produce material transfer which should be taken into account in wear volume calculation [2, 4, 23]. The total net volume V_{tot} of tribological system is typically calculated as:

$$V_{\text{tot}} = V_{\text{disc}} + V_{\text{ball}}$$

$$V_{\text{disc}} = (V_{\text{disc}^-}) - (V_{\text{disc}^+})$$

$$V_{\text{ball}} = (V_{\text{ball}^-}) - (V_{\text{ball}^+})$$

where positive volumes: (V_{disc^+}) and (V_{ball^+}) were considered as transferred materials or build-up while negative volumes (V_{disc^-}) and (V_{ball^-}) as lost material. Figure 2.4 presents an example of volumetric wear calculation.

Fretting wear is observed in various industries—it can be a critical issue in aerospace, automotive, nuclear and biomedical components [15, 16].

To minimise fretting wear, various solutions are taken into consideration—they include selection of optimum contact geometry, surface treatment (like shot peening), the deposition of low friction coatings and the development of new materials.

Previous research reported in [23] suggested that there are about 50 variables affecting the fretting wear process. However, typical experimental conditions in fretting studies include frequency, contact pressure (normal load) and relative displacement.

Fig. 2.4 Example of volumetric wear calculation: volumetric wear of ball:
$(V_{ball^-}) = 3,070,928\ \mu m^3$,
$(V_{ball^+}) = 1951.22\ \mu m^3$,
$V_{ball} = 3,068,976.78\ \mu m^3$
(**a**), volumetric wear of disc:
$(V_{disc^-}) = 2,150,628\ \mu m^3$,
$(V_{disc^+}) = 8352.23\ \mu m^3$,
$V_{disc} = 2,142,275.77\ \mu m^3$
(**b**), total volumetric wear:
$V_{total} = 5,211,252.55\ \mu m^3$

(a)

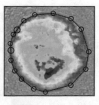

	Hole	Peak
Surface (mm²)	0.391517	0.00533515
Volume (µm³)	3070928	1951.22
Max. depth/height (µm)	21.7993	1.52468
Mean depth/height (µm)	7.84367	0.365730

(b)

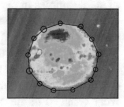

	Hole	Peak
Surface (mm²)	0.409176	0.00858805
Volume (µm³)	2150686	8352.23
Max. depth/height (µm)	15.6712	5.05131
Mean depth/height (µm)	5.25614	0.972541

The presence of dimples on one sliding surface affects wear during dry fretting. The authors of paper [48] analysed various roles of oxide wear debris presence during dry gross slip by allowing them to escape from the contact zone during sliding into oil pockets. Their presence in the interface protected surfaces for dominant adhesion and harmed surfaces when abrasion was a main type of wear. Similar finding were obtained in other studies [26, 27].

Lubrication can be one of the methods for preventing fretting damage [14, 31, 43, 52, 58]. Palliative effect against fretting wear is important in the gross slip regime, compared to the dry condition [51].

2.2 The Effects of Operating Parameters on Fretting

The typical operating parameters are: amplitude, normal load and frequency. In the dry gross slip regime wear increased proportionally to amplitude of displacement [13, 29, 34, 55]. It was found that an increase in the displacement amplitude led also to growth of the coefficient of friction. However, works [8, 25] revealed that an amplitude increase had a small influence on the resistance to motion, although it caused an increase in wear. The authors of papers [29, 50] showed that the change in the amplitude of oscillation caused the change of a dominant wear mechanism.

A reduction of the normal load typically decreased the wear volume [1, 41]. However, an increase in the contact pressure can reduce sliding amplitude leading to operating in the partial slip regime [40], causing a reduction in generation of wear debris.

The effect of a frequency on fretting was also studied by many researchers. Toth [45] found that fretting wear of steel specimens decreased with increasing frequency; frequencies were in the range: 0.01–60 Hz. Soderberg et al. [44] revealed that in the gross slip fretting wear was marginally affected by the frequency. The authors of paper [33] found that displacement amplitude at the transition between the partial slip and the gross slip was reduced with the frequency smaller than 300 Hz. Li et al. [30] studied experimentally the effect of the frequency (between 1 and 80 Hz) on contact of steel samples in dry fretting conditions. Increasing frequency caused a reduction of the accumulated dissipated energy and an increase in the maximum friction coefficient. Park et al. [36, 37] studied a fretting contact between tin-plated and tin-coated brass and found that the wear depth increased with increasing frequency. Fretting wear of sintered iron was experimentally studied [22]. Higher friction coefficients coincided with smaller sliding amplitude and higher frequency, while the amount of wear changed in the opposite tendency. The wear performance of two magnesium alloys was investigated under dry fretting conditions [9]. Volumetric wear decreased with increasing frequency and increased with increasing normal load, and, however, the fretting frequency had marginal effect on the friction coefficient. In only a few works, the influence of the frequency on the coefficient of friction was investigated.

2.3   The Effect of Surface Texture on Fretting

The surface treatment is essential to reduce a damage caused by fretting; however, effects of the surface topography are mostly neglected. However, the surface topography seems to have huge impact on debris presence in the contact zone, since there is opinion that on a rough surface abrasive particles can escape from contact zone to neighbouring valleys, instead of ploughing. Therefore, increasing the surface texture height seems to be an approach of improving the fretting resistance [7, 48]. However, sometimes the opposite results can be obtained. Kubiak et al. [17, 18] achieved smaller friction for rougher disc textures. However, Pawlus et al. [38] found that for rougher disc topographies, the friction force and wear of the tribological system increased. The effect of sphere surface texture on dry fretting in the gross slip regime was analysed experimentally by Yoon et al. [53]. Lower values of maximum friction coefficient in the initial part of test were observed for smoother sphere than for the rougher sphere. Li et al. [30] found that at the frequency of 80 Hz the accumulated dissipated energy for assembly with smoother polished sphere was larger than that with the rougher non-polished sphere. Raeymaekers and Talke [40] observed that the total dissipated energy was slightly higher for the laser polished hemisphere than for the regular hemisphere. However, the authors of papers [30, 40] only estimated wear values. The authors of [21] found that rougher coating coatings had much deeper wear scars than smoother coating. Cho et al. [3] revealed that a higher fretting wear resistance was attributed to an increase in the surface roughness height. When steel disc of various surface roughness co-acted with ceramic ball, both wear and the coefficient of friction were inversely proportional to disc roughness height [24]. The surface texture affected the friction force at the transition between partial slip and gross slip; for smaller roughness height, the coefficient of friction was larger [19, 20]. The authors of paper [28] after fretting wear modelling found that a rougher surface led to higher wear. Pereira et al. [39] obtained a negligible effect of surface roughness on fretting wear. Lu et al. [32] analysed the effect of surface roughness on torsional fretting. They found that the position of the one-directional structure with regard to direction of motion substantially affects fretting wear.

In lubricated fretting, rough surface can help to store oil; therefore, the friction and wear may be decreased for surfaces of a big amplitude. This assumption was confirmed experimentally [19, 42]. In these conditions, wear can be reduced by surface texturing [10, 12, 35]

References

1. F. Akhtar, S.J. Guo, Microstructure, mechanical and fretting wear properties of TiC-stainless steel composites. Mater. Charact. **59**, 84–90 (2008). https://doi.org/10.1016/j.matchar.2006.10.021
2. K.C. Budinsky, Effect of hardness differential on metal-to-metal fretting damage. Wear **301**, 501–507 (2013). https://doi.org/10.1016/j.wear.2013.01.003
3. I.S. Cho, A. Amanov, D.H. Kwak et al., The influence of surface modification techniques on fretting wear of Al–Si alloy prepared by gravity die casting. Mater. Des. **65**, 401–409 (2015). https://doi.org/10.1016/j.matdes.2014.09.036
4. K. Elleuch, S. Fouvry, Wear analysis of A357 aluminum alloy under fretting. Wear **253**, 662–672 (2002). https://doi.org/10.1016/S0043-1648(02)00116-3
5. S. Fouvry, Ph Kapsa, L. Vincent, Analysis of sliding behavior for fretting loading: determination of transition criteria. Wear **185**, 21–46 (1995). https://doi.org/10.1016/0043-1648(94)06582-9
6. S. Fouvry, Ph Kapsa, L. Vincent, Quantification of fretting damage. Wear **200**, 186–205 (1996). https://doi.org/10.1016/S0043-1648(96)07306-1
7. Y. Fu, J. Wei, A.W. Batchelor, Some considerations on the mitigation of fretting damage by the application of surface-modification technologies. J. Mater. Process. Technol. **99**, 231–245 (2000). https://doi.org/10.1016/S0924-0136(99)00429-X
8. Q. Hu, I.R. McColl, S.J. Harris et al., The role of debris in the fretting wear of a SiC reinforced aluminium alloy matrix composite. Wear **245**, 10–21 (2000). https://doi.org/10.1016/S0043-1648(00)00461-0
9. W. Huang, B. Hou, Y. Pang et al., Fretting wear behavior of AZ91D and AM60B magnesium alloy. Wear **260**, 1173–1178 (2006). https://doi.org/10.1016/j.wear.2005.07.023
10. M. Imai, H. Teramoto, Y. Shimauchi et al., Effect of oil supply on fretting wear. Wear **110**, 217–225 (1986). https://doi.org/10.1016/0043-1648(86)90099-2
11. A. Iwabuchi, The role of oxide particles in the fretting wear of mild steel. Wear **151**, 337–344 (1991). https://doi.org/10.1016/0043-1648(91)90257-U

12. T. Jibiki, M. Shima, T. Motoda et al., Role of surface micro-texturing in acceleration of initial running-in during lubricated fretting. Tribol. Online **5**, 33–39 (2010). https://doi.org/10.2474/trol.5.33

13. M. Kalin, J. Vizintin, S. Novak, Effect of fretting conditions on the wear of silicon nitride against bearing steel. Mater. Sci. Eng. **A220**, 191–199 (1996). https://doi.org/10.1016/S0921-5093(96)10457-3

14. M. Kalin, J. Vizintin, S. Novak et al., Wear mechanisms in oil-lubricated and dry fretting of silicon nitride against bearing steel contacts. Wear **210**, 27–38 (1997). https://doi.org/10.1016/S0043-1648(97)00082-3

15. H.K. Ki, Y.H. Lee, S.P. Heo, Mechanical and experimental investigation on nuclear fuel fretting. Tribol. Int. **39**, 1305–1319 (2006). https://doi.org/10.1016/j.triboint.2006.02.027

16. K. Kim, Fretting studies on electroplated brass contacts. Int. J. Mech. Sci. **140**, 306–312 (2018). https://doi.org/10.1016/j.ijmecsci.2018.03.012

17. K.J. Kubiak, M. Bigerelle, T.G. Mathia et al., Roughness of interface in dry contact under fretting conditions, in *Proceedings on the 13th International Conference on Metrology and Properties of Engineering Surfaces*, Twickenham Stadium, UK, 12–15 April 2011

18. K.J. Kubiak, T.E. Liskiewicz, T.G. Mathia, Surface morphology in engineering applications: Influence of roughness on sliding and wear in dry fretting. Tribol. Int. **44**, 1427–1432 (2011). https://doi.org/10.1016/j.triboint.2011.04.020

19. K.J. Kubiak, T.G. Mathia, Influence of roughness on contact interface in fretting under dry and boundary lubricated sliding regimes. Wear **267**, 315–321 (2009). https://doi.org/10.1016/j.wear.2009.02.011

20. K.J. Kubiak, T.G. Mathia, S. Fouvry, Interface roughness effect on friction map under fretting contact conditions. Tribol. Int. **43**, 1500–1507 (2010). https://doi.org/10.1016/j.triboint.2010.02.010

21. L. Lee, E. Regis, S. Descartes et al., Fretting wear behavior of Zn–Ni alloy coatings. Wear **330–331**, 112–121 (2015). https://doi.org/10.1016/j.wear.2015.02.043

22. E.R. Leheup, D. Zhang, J.R. Moon, Fretting wear of sintered iron under low normal pressure. Wear **221**, 86–92 (1998). https://doi.org/10.1016/S0043-1648(98)00265-8

23. J.D. Lemm, A.R. Warmuth, S.R. Pearson et al., The influence of surface hardness on the fretting wear of steel pairs—its role in debris retention in contact. Tribol. Int. **81**, 258–266 (2015). https://doi.org/10.1016/j.triboint.2014.09.003

24. A. Lenart, P. Pawlus, A. Dzierwa, The effect of steel disc surface texture in contact with ceramic ball on friction and wear in dry fretting. Surf. Topogr. Metrol. Prop. **6**(3), 034004 (2018). https://doi.org/10.1088/2051-672X/aac1a2

25. A. Lenart, P. Pawlus, A. Dzierwa et al., The effect of surface topography on dry fretting in the gross slip regime. Arch. Civ. Mech. Eng. **17**(4), 894–904 (2017). https://doi.org/10.1016/j.acme.2017.03.008

26. A. Lenart, P. Pawlus, A. Dzierwa et al., The effect of wear debris removal on the fretting of rough surfaces. Tribologia **4**, 47–54 (2017). https://doi.org/10.5604/01.3001.0010.5993

27. A. Lenart, P. Pawlus, S. Wos et al., The effect of surface texturing on dry gross fretting. Lubricants **6**, 92 (2018). https://doi.org/10.3390/lubricants6040092

28. B.D. Leonard, F. Sadeghi, S. Shinde et al., Rough surface and damage mechanics under wear modeling using the combined finite-discrete element method. Wear **305**, 312–321 (2013). https://doi.org/10.1016/j.wear.2012.12.034

29. J. Li, Y.H. Lu, Effects of displacement amplitude on fretting wear behaviors and mechanism of Inconel 600 Alloy. Wear **304**, 223–230 (2013). https://doi.org/10.1016/j.wear.2013.04.027

30. L. Li, I. Etsion, F.E. Talke, The effect of frequency on fretting in a micro-spherical contact. Wear **270**, 857–865 (2011). https://doi.org/10.1016/j.wear.2011.02.014

31. Q.Y. Liu, Z.R. Zhou, Effect of displacement amplitude in oil-lubricated fretting. Wear **239**, 237–243 (2000). https://doi.org/10.1016/S0043-1648(00)00323-9

32. W. Lu, P. Zhang, X. Liu et al., Influence of surface topography on torsional fretting under flat-on-flat contact. Tribol. Int. **109**, 367–372 (2017). https://doi.org/10.1016/j.triboint.2017.01.001

33. M. Odfalk, O. Vingsbo, Influence of normal force and frequency on fretting. Tribol. Trans. **33**, 604–610 (1990). https://doi.org/10.1080/10402009008981995

34. N. Ohmae, T. Tsukizoe, The effect of slip amplitude on fretting. Wear **27**, 281–294 (1974). https://doi.org/10.1016/0043-1648(74)90114-8

35. M. Okamoto, T. Jibiki, S. Ito et al., Role of cross-grooved type texturing in acceleration of initial running-in under lubricated fretting. Tribol. Int. **100**, 126–131 (2016). https://doi.org/10.1016/j.triboint.2015.12.012

36. Y.W. Park, G.R. Bapu, K.Y. Lee, Studies of tin coated brass contacts in fretting conditions under different loads and frequencies. Surf. Coat. Technol. **201**, 7939–7951 (2007). https://doi.org/10.1016/j.surfcoat.2007.03.039

37. Y,W. Park, T.S. Narayanan, K.Y. Lee, Effect of fretting amplitude and frequency on the fretting corrosion behavior of thin plated contacts. Surf. Coat. Technol. **201**, 2181–2192 (2006). https://doi.org/10.1016/j.surfcoat.2006.03.031

38. P. Pawlus, R. Michalczewski, A. Lenart et al., The effect of random surface topography height on fretting in dry gross slip conditions. Proc. Inst. Mech. Eng. J J. Eng. **228**, 1374–1391 (2014). https://doi.org/10.1177/1350650114539467

39. K. Pereira, T. Yue, M.A. Wahab, Multiscale analysis of the effect of roughness on fretting wear. Tribol. Int. **110**, 222–231 (2017). https://doi.org/10.1016/j.triboint.2017.02.024

40. B. Raeymaekers, F.E. Talke, The effect of laser polishing on fretting wear between a hemisphere and a flat plate. Wear **269**, 416–423 (2010). https://doi.org/10.1016/j.wear.2010.04.027

41. R. Ramesh, R. Gnanamoorthy, Effect of hardness on fretting behavior of structural steel. Mater. Des. **28**, 1447–1452 (2007). https://doi.org/10.1016/j.matdes.2006.03.020

42. J. Sato, M. Shima, T. Sugawara et al., Effect of lubricants on fretting wear of steel. Wear **125**, 83–95 (1988). https://doi.org/10.1016/0043-1648(88)90195-0

43. M. Shima, H. Suetake, I.R. McColl et al., On the behaviour of an oil lubricated fretting contact. Wear **210**, 304–310 (1997). https://doi.org/10.1016/S0043-1648(97)00078-1

44. S. Soderberg, U. Bryggman, T. McCullough, Frequency effects in fretting wear. Wear **110**, 19–34 (1986). https://doi.org/10.1016/0043-1648(86)90149-3

45. L. Toth, The investigation of the steady stage of steel fretting. Wear **20**, 277–286 (1972). https://doi.org/10.1016/0043-1648(72)90409-7

46. M. Varenberg, I. Etsion, E. Altus, Theoretical substantiation of the slip index approach to fretting. Tribol. Lett. **19**, 263–264 (2005). https://doi.org/10.1007/s11249-005-7442-8

47. M. Varenberg, I. Etsion, G. Halperin, Slip index: a new unified approach to fretting. Tribol. Lett. **17**(3), 569–573 (2004). https://doi.org/10.1023/B:TRIL.0000044506.98760.f9

48. M. Varenberg, G. Halperin, I. Etsion, Different aspects of the role of wear debris in fretting wear. Wear **252**, 902–910 (2002). https://doi.org/10.1016/S0043-1648(02)00044-3

49. O. Vingsbo, S. Soerberg, On fretting maps. Wear **126**, 131–147 (1988). https://doi.org/10.1016/0043-1648(88)90134-2

50. D. Wang, D. Zhang, S. Ge, Effect of displacement amplitude on fretting fatigue behavior of hoisting rope wires in low cycle fatigure. Tribol. Int. **52**, 178–189 (2012). https://doi.org/10.1016/j.triboint.2012.04.008

51. Z.A. Wang, Z.R. Zhou, An investigation of fretting behaviour of several synthetic base oils. Wear **267**, 1399–1404 (2009). https://doi.org/10.1016/j.wear.2008.12.092

52. A.R. Warmuth, W. Sun, P.H. Shipway, The roles of contact conformity, temperature and displacement amplitude on the lubricated fretting wear of a steel-on-steel contact. R. Soc. Open Sci. **3**, 150637 (2016). https://doi.org/10.1098/rsos.150637

53. Y. Yoon, I. Etsion, F.E. Talke, The evolution of fretting wear in a micro-spherical contact. Wear **270**, 567–575 (2011). https://doi.org/10.1016/j.wear.2011.01.013

54. J. Yu, L. Qian, B. Yu et al., Nanofretting behavior of monocrystalline silicon (1 0 0) against SiO$_2$ microsphere in vacuum. Tribol. Lett. **34**, 31–40 (2009). https://doi.org/10.1007/s11249-008-9385-3

55. Z.R. Zhou, E. Sauger, J.J. Liu et al., Nucleation and early growth of tribologically transformed structure (TTS) induced by fretting. Wear **212**, 50–58 (1997). https://doi.org/10.1016/S0043-1648(97)00141-5

56. Z.R. Zhou, S. Nakazawa, M.H. Zhu et al., Progress in fretting maps. Tribol. Int. **39**, 1068–1073 (2006). https://doi.org/10.1016/j.triboint.2006.02.001

57. Z.R. Zhou, L. Vincent, Mixed fretting regimes. Wear **181–183**, 531–536 (1995). https://doi.org/10.1016/0043-1648(95)90168-X

58. Z.R. Zhou, L. Vincent, Lubrication in fretting—a review. Wear **225–229**, 962–967 (1999). https://doi.org/10.1016/S0043-1648(99)00038-1

Chapter 3
The Effect of Frequency and Normal Load on Dry Gross Fretting of Rough Surfaces

3.1 Experimental Conditions

The typical operating parameters are amplitude, normal load and frequency. In previous research reported in [1], it was found that the wear volume was proportional to the amplitude to displacement; however, the effect of this amplitude on the coefficient of friction was marginal. These investigations were carried out using the same tribotester and specimen-counter specimen configuration as in research reported here. The normal load was 45 N, the frequency was 20 Hz, the number of cycles was 18,000, the test duration was 15 min, while the amplitudes of oscillation were 0.1, 0.075 and 0.05 mm proportional. Therefore, we studied only effects of the normal force and the frequency of oscillations on dry gross fretting of rough surfaces.

The main objective of the research was to determine the effect of the surface topography of steel discs on the frictional resistance and wear under dry fretting conditions at a constant displacement amplitude of 0.05 mm. An additional aim was to estimate the effect of the frequency and the normal load on fretting of rough surfaces.

Tribological tests were carried out using an Optimol SRv5 tester in flat-on-ball configuration under the dry gross slip condition. Hardness of steel ball was 60 HRC, while hardness of steel disc was 40 HRC.

A relative humidity was between 25 and 35%. The frequency and the normal load were changed at three levels: 20, 50 and 80 Hz (frequency), and 15, 45 and 75 N (load). The tests were carried out in two variants: at a constant test time (15 min) and with a constant number of oscillations (18,000). For a frequency of 50 Hz, the number of oscillations (18,000) corresponded to the test duration of 6 min and for the frequency of 80 Hz to 3 min and 45 s, respectively. In each test condition, the displacement amplitude was smaller than the radius of the elastic contact. The slip index [2, 3] obtained value characteristic for the gross slip regime. Table 3.1 presents the test conditions.

© The Author(s), under exclusive license to Springer Nature Switzerland AG 2020
P. Pawlus et al., *Dry Gross Fretting of Rough Surfaces*, Manufacturing
and Surface Engineering, https://doi.org/10.1007/978-3-030-31563-4_3

Table 3.1 Parameters of tribological tests

Number of test	Frequency f, Hz	Normal load P, N	Time t, min
A	20	15	15
B	50	15	15
C	50	15	6
D	80	15	15
E	80	15	3 min 45 s
F	20	45	15
G	50	45	15
H	50	45	6
I	80	45	15
J	80	45	3 min 45 s
K	20	75	15
L1	50	75	15
M	50	75	6
N	80	75	15
O	80	75	3 min 45 s

Nine types of disc surfaces were selected for this study. These were samples with one-process surfaces.

It was decided to not take into consideration two-process surfaces because it was confirmed in previous research [1] that the additional treatment did not improve tribological properties of frictional pairs.

Six parameters were selected to completely describe two-process surfaces. The Sq parameter (standard deviation of height) characterises amplitude, the skewness Ssk and the kurtosis Sku the shape of the ordinate distribution, Sal and Str spatial however Sdq hybrid surface properties. These parameters were defined in Chap. 2. The values of the parameters: Sq, Ssk, Sku, Sal and Str are required to model anisotropic one-process surface [4]. The slope Sdq was added because it characterises both amplitude and spatial surface properties.

Table 3.2 presents surface topography parameters of tested disc samples, while Fig. 3.1 shows contour plots of the disc surfaces. Surface topographies were measured by a white light interferometer Talysurf CCI Lite of 0.01 vertical resolution. The measured area was 3.3 mm × 3.3 mm (1024 × 1024 points). Surfaces after measurement were only levelled, and digital filtration was not used. If necessary, spikes were excluded by surface truncation.

The sample after polishing P was characterised by the smallest roughness height and the largest correlation length. The lapped sample L was also smooth with the smallest correlation length. They were one-process random surfaces of symmetric ordinate distribution, and the Str parameter was between 0.14 and 0.19 (anisotropic or mixed surfaces). Ground one-process random surfaces G1 and G2 were characterised by a negative value of the skewness Ssk, which is characteristic feature of

Table 3.2 Surface topography parameters of tested disc samples

Surface	Sq, μm	Ssk	Sku	Sal, mm	Str	Sdq	Type of finishing treatment
P	0.022	−0.01	3.2	0.13	0.194	0.0036	Polishing
L	0.045	−0.08	3.9	0.0061	0.139	0.0162	Lapping
G1	1.19	−0.84	4.4	0.064	0.032	0.183	Grinding
G2	0.3	−0.61	3.8	0.034	0.255	0.066	Grinding
M1	0.32	−0.11	2.7	0.101	0.07	0.03	Milling
M2	0.59	0.18	2.28	0.05	0.02	0.089	Milling
M3	9.4	−0.8	3.6	0.11	0.15	0.36	Milling
VB1	1.22	−0.6	5.7	0.009	0.77	0.35	Vapour blasting
VB2	1.9	−0.6	4.1	0.013	0.8	0.46	Vapour blasting

textures after grinding. The G1 surface was rougher, while G2 smoother. They were anisotropic, one-directional textures, although the Str parameter of G2 surface was 0.255. Surfaces after milling had periodic structure. Smooth surfaces M1 and M2 were anisotropic one-directional structures of asymmetric ordinate distribution. The amplitude of M3 surface was the highest from all analysed samples. Samples after vapour blasting VB1 and VB2 were isotropic rough surfaces. One can see that the tested surface were characterised by various height (the Sq parameter was between 0.0022 and 9.4 μm. The range of the correlation length was also wide (between 0.006 and 0.13 mm). They were random or periodic textures.

The roughness height of a ball sample characterised by the Ra parameter was 0.14 μm.

During tests, the main directions of one-directional anisotropic surfaces (lays) were positioned perpendicularly to the direction of ball motion.

For the test duration of 6 min, the average value of the friction coefficient was determined between 50 and 360 s and the maximum value between 300 and 360 s. When test duration was 3 min and 45 s, the values of the coefficient of friction were obtained between 50 and 225 s (average value) and between 200 and 225 s (maximum value), respectively.

The number of test repetitions for the same assembly was at least 3.

3.2 Results and Discussion

When the normal load was 75 N, the diameter of the elastic contact was 0.27 mm, while the elastic contact pressure was 1961 MPa. For the load of 15 N, the elastic contact pressure was 1147 MPa and the diameter of the elastic contact was 0.158 mm, this diameter was smaller than the stroke of 0.1 mm—therefore, the gross slip occurred.

Tables 3.3, 3.4, 3.5, 3.6, 3.7, 3.8, 3.9, 3.10, 3.11, 3.12, 3.13, 3.14, 3.15, 3.16 and 3.17 list the results of the tribological tests for various conditions from A to O. Apart

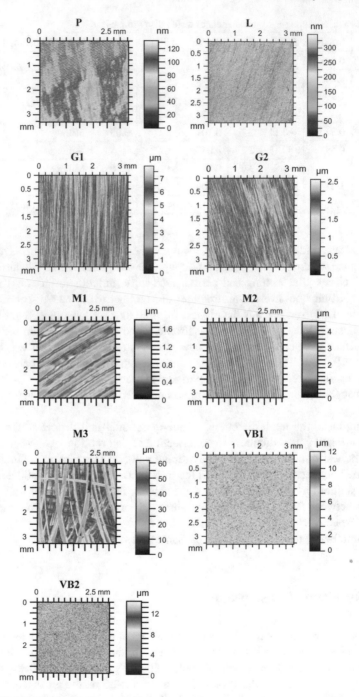

Fig. 3.1 Contour plots of selected disc samples

Table 3.3 Results of tests A for the following test conditions: $f = 20$ Hz, $P = 15$ N, $t = 15$ min; volumetric wear of ball V_{ball}, volumetric wear of disc V_{disc}, the total wear volume of the tribological system V_{total}, the average coefficient of friction μ_{50-900} and the final coefficient of friction $\mu_{600-900}$

Surface	V_{ball}, μm^3	V_{disc}, μm^3	V_{total}, μm^3	μ_{50-900}	$\mu_{600-900}$	Sq, μm
P	2,664,075	1,124,182	3,788,257	1.121	1.13	0.02
L	2,415,922	1,163,221	3,579,143	1.092	1.098	0.045
G1	2,350,236	681,559.5	3,031,796	1.07	1.104	1.19
G2	2,868,609	668,730	3,537,339	1.092	1.101	0.3
M1	1,797,679	1,046,566	2,844,245	1.033	1.04	0.32
M2	2,166,975	879,315.2	3,046,290	1.03	1.06	0.59
M3	2,296,975	870,296	3,167,271	1.092	1.105	9.4
VB1	2,511,157	1,441,374	3,952,531	1.14	1.175	1.22
VB2	1,869,126	2,087,947	3,957,073	1.09	1.09	1.1

Table 3.4 Results of tests B for the following test conditions: $f = 20$ Hz, $P = 15$ N, $t = 15$ min; volumetric wear of ball V_{ball}, volumetric wear of disc V_{disc}, the total wear volume of the tribological system V_{total}, the average coefficient of friction μ_{50-900} and the final coefficient of friction $\mu_{600-900}$

Surface	V_{ball}, μm^3	V_{disc}, μm^3	V_{total}, μm^3	μ_{50-900}	$\mu_{600-900}$	Sq, μm
P	5,797,751	2,431,175	8,228,926	1.246	1.246	0.02
L	8,451,657	316,081.5	8,767,738	1.14	1.178	0.045
G1	8,077,369	4,785,58.5	8,555,928	1.14	1.18	1.19
G2	7,927,202	5,761,40.5	8,503,343	1.27	1.285	0.3
M1	6,337,855	1,929,885	8,267,740	1.09	1.1	0.32
M2	8,169,651	7,383,36.7	8,907,987	1.11	1.135	0.59
M3	6,478,331	2,172,503	8,650,834	1.208	1.295	9.4
VB1	8,097,069	626,159	8,723,228	1.16	1.16	1.22
VB2	6,171,737	2,615,549	8,787,286	1.18	1.18	1.1

Table 3.5 Results of tests C for the following test conditions: $f = 20$ Hz, $P = 15$ N, $t = 6$ min); volumetric wear of ball V_{ball}, volumetric wear of disc V_{disc}, the total wear volume of the tribological system V_{total}, the average coefficient of friction μ_{50-360} and the final coefficient of friction $\mu_{300-360}$

Surface	V_{ball}, μm^3	V_{disc}, μm^3	V_{total}, μm^3	μ_{50-900}	$\mu_{600-900}$	Sq, μm
P	3,710,031	−884,456	2,825,575	1.195	1.205	0.02
L	3,229,007	−677,278	2,551,729	1.176	1.208	0.045
G1	3,194,325	−668,938	2,525,387	1.13	1.15	1.19
G2	3,127,244	−665,225	2,462,019	1.208	1.213	0.3
M1	4,378,277	−780,502	3,597,775	1.085	1.105	0.32
M2	4,095,904	−352,505	3,743,399	1.09	1.11	0.59
M3	4,991,932	−1,831,987	3,159,945	1.137	1.14	9.4
VB1	4,060,742	47,615	4,108,357	1.163	1.169	1.22
VB2	4,002,914	−33,119	3,969,795	1.163	1.179	1.1

Table 3.6 Results of tests D for the following test conditions: $f = 20$ Hz, $P = 15$ N, $t = 15$ min; volumetric wear of ball V_{ball}, volumetric wear of disc V_{disc}, the total wear volume of the tribological system V_{total}, the average coefficient of friction μ_{50-900} and the final coefficient of friction $\mu_{600-900}$

Surface	V_{ball}, μm^3	V_{disc}, μm^3	V_{total}, μm^3	μ_{50-900}	$\mu_{600-900}$	Sq, μm
P	14,282,270	412,170	14,694,440	1.298	1.259	0.02
L	12,701,288	1,031,080	13,732,368	1.28	1.234	0.045
G1	12,406,512	577,415	12,983,927	1.28	1.23	1.19
G2	14,243,605	695,752	14,939,357	1.285	1.25	0.3
M1	12,763,414	816,961	13,580,375	1.22	1.22	0.32
M2	13,596,240	857,955	14,454,195	1.23	1.22	0.59
M3	10,045,590	4,586,161	14,631,751	1.324	1.362	9.4
VB1	15,344,241	276,963	15,621,204	1.285	1.246	1.22
VB2	11,674,451	2,821,416	14,495,867	1.285	1.259	1.1

Table 3.7 Results of tests E for the following test conditions: $f = 80$ Hz, $P = 15$ N, $t = 3$ min 45 s; volumetric wear of ball V_{ball}, volumetric wear of disc V_{disc}, the total wear volume of the tribological system V_{total}, the average coefficient of friction μ_{50-225} and the final coefficient of friction $\mu_{200-225}$

Surface	V_{ball}, μm^3	V_{disc}, μm^3	V_{total}, μm^3	μ_{50-900}	$\mu_{600-900}$	Sq, μm
P	2,995,334	−877,504	2,117,830	1.298	1.326	0.02
L	2,999,417	−771,521	2,227,896	1.298	1.324	0.045
G1	3,030,008	−825,213	2,204,795	1.246	1.295	1.19
G2	3,801,098	−1,360,373	2,440,725	1.31	1.342	0.3
M1	2,957,066	−105,528	2,851,538	1.22	1.255	0.32
M2	2,835,054	92,874	2,927,928	1.225	1.26	0.59
M3	3,166,291	90,927	3,257,218	1.304	1.36	9.4
VB1	3,252,566	50,842	3,303,408	1.311	1.33	1.22
VB2	2,554,635	233,408	2,788,043	1.311	1.343	1.1

Table 3.8 Results of tests F for the following test conditions: $f = 20$ Hz, $P = 45$ N, $t = 15$ min; volumetric wear of ball V_{ball}, volumetric wear of disc V_{disc}, the total wear volume of the tribological system V_{total}, the average coefficient of friction μ_{50-900} and the final coefficient of friction $\mu_{600-900}$

Surface	V_{ball}, μm^3	V_{disc}, μm^3	V_{total}, μm^3	μ_{50-900}	$\mu_{600-900}$	Sq, μm
P	3,373,302	1,904,083	5,277,385	0.93	0.99	0.02
L	3,787,737	757,194	4,544,931	0.93	0.98	0.045
G1	3,233,224	1,071,328	4,304,552	0.90	0.96	1.19
G2	3,590,540	531,382	4,121,922	0.91	0.98	0.3
M1	3,407,793	219,212	3,627,005	0.86	0.92	0.32
M2	3,069,400	76,333	3,145,733	0.85	0.92	0.59
M3	2,721,160	1,173,267	3,894,427	0.92	0.98	9.4
VB1	4,252,367	189,999	4,442,366	0.93	0.97	1.22
VB2	2,090,441	2,339,596	4,430,037	0.92	0.98	1.1

Table 3.9 Results of tests G for the following test conditions: $f = 50$ Hz, $P = 45$ N, $t = 15$ min; volumetric wear of ball V_{ball}, volumetric wear of disc V_{disc}, the total wear volume of the tribological system V_{total}, the average coefficient of friction μ_{50-900} and the final coefficient of friction $\mu_{600-900}$

Surface	V_{ball}, μm^3	V_{disc}, μm^3	V_{total}, μm^3	μ_{50-900}	$\mu_{600-900}$	Sq, μm
P	10,344,337	579,288	10,923,625	0.968	1.023	0.02
L	10,462,766	1,595,351	12,058,117	0.979	1.012	0.045
G1	11,687,494	−267,961	11,419,533	0.99	1.045	1.19
G2	9,372,638	−985,437	8,387,201	0.991	1.056	0.3
M1	9,612,899	6914	9,619,813	0.89,744	0.97	0.32
M2	7,009,067	2,419,452	9,428,519	0.912	0.9861	0.59
M3	9,823,428	2,127,629	11,951,057	0.99	1.045	9.4
VB1	9,457,838	- 660,512	8,797,326	0.99	1.056	1.22
VB2	9,060,320	- 164,298	8,896,022	0.978	1.045	1.1

Table 3.10 Results of tests H for the following test conditions: $f = 50$ Hz, $P = 45$ N, $t = 6$ min); volumetric wear of ball V_{ball}, volumetric wear of disc V_{disc}, the total wear volume of the tribological system V_{total}, the average coefficient of friction μ_{50-360} and the final coefficient of friction $\mu_{300-360}$

Surface	V_{ball}, μm^3	V_{disc}, μm^3	V_{total}, μm^3	μ_{50-900}	$\mu_{600-900}$	Sq, μm
P	3,639,705	399,832	4,039,537	0.91	0.985	0.02
L	4,101,003	−145,898	3,955,105	0.9075	0.99	0.045
G1	4,016,210	−385,100	3,631,110	0.914	0.995	1.19
G2	4,053,804	−604,408	3,449,396	0.9075	0.99	0.3
M1	2,499,125	263,894	2,763,019	0.84,632	0.934	0.32
M2	3,183,106	−242,814	2,940,292	0.8094	0.912	0.59
M3	4,976,503	−842,274	4,134,229	0.9075	1.01	9.4
VB1	4,411,828	−227,324	4,184,504	0.885	0.97	1.22
VB2	3,128,690	116,557	3,245,247	0.8855	0.96	1.1

Table 3.11 Results of tests I for the following test conditions: $f = 80$ Hz, $P = 45$ N, $t = 15$ min; volumetric wear of ball V_{ball}, volumetric wear of disc V_{disc}, the total wear volume of the tribological system V_{total}, the average coefficient of friction μ_{50-900} and the final coefficient of friction $\mu_{600-900}$

Surface	V_{ball}, μm^3	V_{disc}, μm^3	V_{total}, μm^3	μ_{50-900}	$\mu_{600-900}$	Sq, μm
P	17,750,933	−2,198,131	15,552,802	1.067	1.114	0.02
L	17,040,655	−3,198,103	13,842,552	1.034	1.09	0.045
G1	17,423,567	−3,509,922	13,913,645	1.045	1.078	1.19
G2	15,660,821	−1,893,363	13,767,458	1.034	1.089	0.3
M1	14,437,523	−696,974	13,740,549	0.966	1.045	0.32
M2	15,297,033	−1,589,728	13,707,305	0.911	1.01	0.59
M3	14,384,373	524,599	14,908,972	1.012	1.089	9.4
VB1	14,713,957	−2,128,635	12,585,322	1.051	1.111	1.22
VB2	13,135,898	359,649.5	13,495,548	1.03	1.11	1.1

Table 3.12 Results of tests J for the following test conditions: $f = 80$ Hz, $P = 45$ N, $t = 3$ min 45 s; volumetric wear of ball V_{ball}, volumetric wear of disc V_{disc}, the total wear volume of the tribological system V_{total}, the average coefficient of friction μ_{50-225} and the final coefficient of friction $\mu_{200-225}$

Surface	V_{ball}, μm^3	V_{disc}, μm^3	V_{total}, μm^3	μ_{50-900}	$\mu_{600-900}$	Sq, μm
P	3,195,787	247,408	3,443,195	0.925	0.997	0.02
L	3,089,758	−46,577	3,043,181	0.893	0.978	0.045
G1	3,348,193	−136,211	3,211,982	0.9	0.96	1.19
G2	3,084,793	−115,009	2,969,784	0.8855	0.961	0.3
M1	2,817,895	−188,436	2,629,459	0.765	0.846	0.32
M2	2,891,244	−92,874	2,798,370	0.744	0.825	0.59
M3	2,767,386	−72,748	2,694,638	0.8415	0.931	9.4
VB1	3,407,692	−693,847	2,713,845	0.891	0.98	1.22
VB2	2,499,966	409,170	2,909,136	0.8635	0.941	1.1

Table 3.13 Results of tests K for the following test conditions: $f = 20$ Hz, $P = 75$ N, $t = 15$ min; volumetric wear of ball V_{ball}, volumetric wear of disc V_{disc}, the total wear volume of the tribological system V_{total}, the average coefficient of friction μ_{50-900} and the final coefficient of friction $\mu_{600-900}$

Surface	V_{ball}, μm^3	V_{disc}, μm^3	V_{total}, μm^3	μ_{50-900}	$\mu_{600-900}$	Sq, μm
P	4,409,939	1,123,401	5,533,340	0.809	0.885	0.02
L	3,568,224	1,595,101	5,163,325	0.811	0.874	0.045
G1	3,814,937	341,884	4,156,821	0.794	0.847	1.19
G2	3,651,073	1,889,179	5,540,252	0.797	0.867	0.3
M1	4,174,970	645,680	4,820,650	0.783	0.842	0.32
M2	4,166,599	690,191	4,856,790	0.775	0.832	0.59
M3	3,989,888	1,270,492	5,260,380	0.822	0.874	9.4
VB1	4,523,457	826,897.3	5,350,354	0.832	0.863	1.22
VB2	3,510,998	1,918,569	5,429,567	0.832	0.884	1.1

Table 3.14 Results of tests L1 for the following test conditions: $f = 50$ Hz, $P = 75$ N, $t = 15$ min; volumetric wear of ball V_{ball}, volumetric wear of disc V_{disc}, the total wear volume of the tribological system V_{total}, the average coefficient of friction μ_{50-900} and the final coefficient of friction $\mu_{600-900}$

Surface	V_{ball}, μm^3	V_{disc}, μm^3	V_{total}, μm^3	μ_{50-900}	$\mu_{600-900}$	Sq, μm
P	8,456,733	2,506,459	10,963,192	0.884	0.957	0.02
L	10,834,329	2,128,084	12,962,413	0.905	0.988	0.045
G1	9,875,361	878,653	10,754,014	0.875	0.968	1.19
G2	9,011,990	1,814,828	10,826,818	0.874	0.967	0.3
M1	9,330,062	−421,623	8,908,439	0.853	0.936	0.32
M2	9,230,447	−32,810	9,197,637	0.852	0.928	0.59
M3	9,420,609	−709,967	8,710,642	0.874	0.947	9.4
VB1	9,324,989	−903,954	8,421,035	0.874	0.952	1.22
VB2	7,071,238	1,468,787	8,540,025	0.865	0.951	1.1

Table 3.15 Results of tests M for the following test conditions: $f = 50$ Hz, $P = 75$ N, $t = 6$ min); volumetric wear of ball V_{ball}, volumetric wear of disc V_{disc}, the total wear volume of the tribological system V_{total}, the average coefficient of friction μ_{50-360} and the final coefficient of friction $\mu_{300-360}$

Surface	V_{ball}, μm^3	V_{disc}, μm^3	V_{total}, μm^3	μ_{50-900}	$\mu_{600-900}$	Sq, μm
P	3,679,153	239,334	3,918,487	0.811	0.875	0.02
L	3,540,386	629,924	4,170,310	0.801	0.9	0.045
G1	3,267,398	414,009	3,681,407	0.762	0.85	1.19
G2	3,353,585	622,610	3,976,195	0.770	0.85	0.3
M1	2,610,671	1,005,276	3,615,947	0.748	0.835	0.32
M2	3,063,503	605,772	3,669,275	0.739	0.83	0.59
M3	3,297,286	522,232	3,819,518	0.791	0.87	9.4
VB1	3,374,493	397,334	3,771,827	0.770	0.85	1.22
VB2	3,222,220	635,093	3,857,313	0.765	0.845	1.1

Table 3.16 Results of tests N for the following test conditions: $f = 50$ 8z, $P = 75$ N, $t = 15$ min; volumetric wear of ball V_{ball}, volumetric wear of disc V_{disc}, the total wear volume of the tribological system V_{total}, the average coefficient of friction μ_{50-900} and the final coefficient of friction $\mu_{600-900}$

Surface	V_{ball}, μm^3	V_{disc}, μm^3	V_{total}, μm^3	μ_{50-900}	$\mu_{600-900}$	Sq, μm
P	14,260,539	1,799,467	16,060,006	0.962	1.023	0.02
L	13,847,919	2,410,135	16,258,054	0.957	1.051	0.045
G1	17,035,068	−1,197,631	15,837,437	0.964	1.04	1.19
G2	12,956,018	1,533,566	14,489,584	0.967	1.051	0.3
M1	12,708,187	1,152,008	13,860,195	0.919	0.993	0.32
M2	12,959,026	107,759	13,066,785	0.898	0.995	0.59
M3	13,349,772	2,071,047	15,420,819	0.947	1.030	9.4
VB1	13,141,249	−130,037	13,011,212	0.947	1.015	1.22
VB2	11,258,933	1,923,046	13,181,979	0.926	0.996	1.1

from the values of wear and friction coefficient, the mean value of the Sq parameter was also given. Examples of curves presenting the maximum coefficient of friction versus time for selected frictional pairs containing L, VB1 and M2 discs are shown in Figs. 3.2, 3.3, 3.4, 3.5, 3.6, 3.7, 3.8, 3.9, 3.10, 3.11, 3.12, 3.13, 3.14, 3.15 and 3.16. The scatter of values of the friction coefficient and wear for friction pairs with the same disc were typically higher for the smaller normal load, for the load of 15 N the scatter of the coefficient of friction did not exceed 0.028, whereas total wear—15%, an increase of the normal load to 45 N caused the scatter of the coefficient of friction smaller than 0.023, and total wear—10%. At the largest value of normal load, the scatter usually did not exceed 0.018 and 7%, respectively.

After the test A, the lowest values of coefficients of friction were obtained for friction pairs with samples after finish milling M1 and M2 (Table 3.3). The small average coefficient of friction was observed for the pairs containing the discs after

Table 3.17 Results of tests O for the following test conditions: $f = 80$ Hz, $P = 75$ N, $t = 3$ min 45 s; volumetric wear of ball V_{ball}, volumetric wear of disc V_{disc}, the total wear volume of the tribological system V_{total}, the average coefficient of friction μ_{50-225} and the final coefficient of friction $\mu_{200-225}$

Surface	V_{ball}, μm^3	V_{disc}, μm^3	V_{total}, μm^3	μ_{50-900}	$\mu_{600-900}$	Sq, μm
P	2,882,556	1,083,315	3,965,871	0.826	0.91	0.02
L	2,471,692	1,163,320	3,635,012	0.816	0.9	0.045
G1	2,638,576	1,028,157	3,666,733	0.780	0.87	1.19
G2	3,239,262	689,140	3,928,402	0.811	0.91	0.3
M1	2,680,121	808,787	3,488,908	0.765	0.86	0.32
M2	1,973,097	1,338,138	3,311,235	0.724	0.79	0.59
M3	3,079,148	312,584	3,391,732	0.785	0.85	9.4
VB1	2,989,938	654,721	3,644,659	0.822	0.9	1.22
VB2	2,489,897	1,114,476	3,604,373	0.822	0.88	1.1

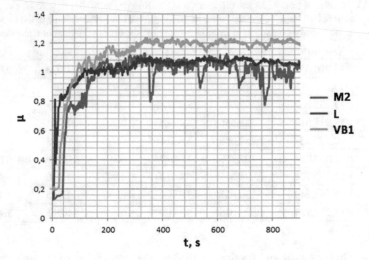

Fig. 3.2 Maximum coefficient of friction versus time for selected frictional pairs, in conditions A

rough grinding G1. In these three cases, the smallest values of total wear of the tribological system were obtained. However, the highest coefficients of friction and large wear were obtained for the frictional pair containing disc after vapour blasting VB1. The coefficient of friction obtained stable values after 350 s (Fig. 3.2). Wear of balls was much higher than the wear of the discs (except for the samples after vapour blasting VB1 and VB2). The coefficient of friction was proportional to total wear (the linear coefficient of correlation was higher than 0.75).

When the frequency of oscillation increased, the lowest coefficients of friction in test B were obtained for samples after finish milling M1 and M2 (Table 3.4). No other tendencies were observed. The coefficient of friction reached its maximum

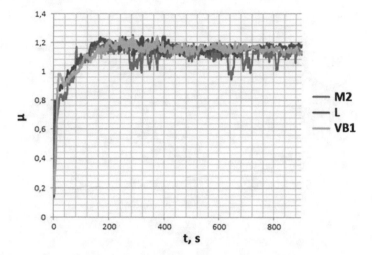

Fig. 3.3 Maximum coefficient of friction versus time for selected frictional pairs, in conditions B

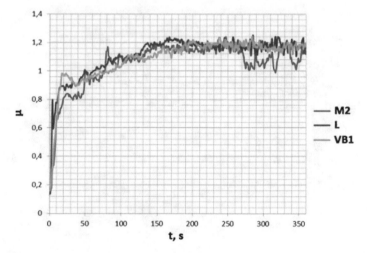

Fig. 3.4 Maximum coefficient of friction versus time for selected frictional pairs, in conditions C

value after about 200 s and then slightly decreased (Fig. 3.3). Total wear of the tested tribological systems was similar. Wear of balls was much higher than wear of discs and inversely correlated with it.

It was noticed during the test C, that the smallest values of the coefficient of friction were again obtained for samples after finish milling M1 and M2 (Table 3.5; Fig. 3.4). Slightly higher frictional resistance was obtained for frictional pairs containing discs after rough milling M3 and rough grinding G1. In most cases, the negative values of disc wear were noticed; only the sample after vapour blasting VB1 was the exception.

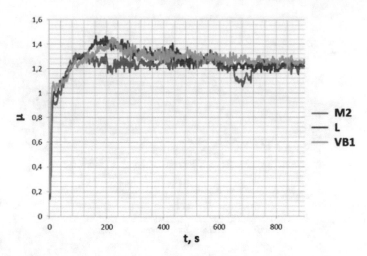

Fig. 3.5 Maximum coefficient of friction versus time for selected frictional pairs, in conditions D

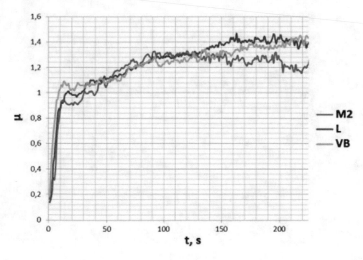

Fig. 3.6 Maximum coefficient of friction versus time for selected frictional pairs, in conditions E

During the test D, the smallest values of the coefficient of friction (Table 3.6) were again observed for friction pairs containing finish-milled samples M1 and M2. The rough-milled disc M3 led to the highest values of the frictional resistance. Similar to the frequency of 50 Hz, the coefficient of friction reached its maximum after about 200 s and then decreased (Fig. 3.5). Wear of balls was much higher than the wear of discs and inversely correlated with it. As in most cases, significant wear of the frictional pair containing the disc after vapour blasting VB1 was observed. The highest values of wear as well as the coefficient of friction were obtained in comparison with previously analysed operating conditions.

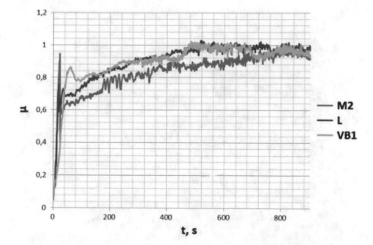

Fig. 3.7 Maximum coefficient of friction versus time for selected frictional pairs, in conditions F

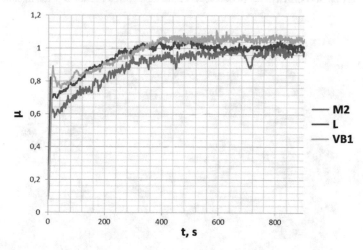

Fig. 3.8 Maximum coefficient of friction versus time for selected frictional pairs, in conditions G

During the test E, small values of the friction coefficient and wear were observed for the frictional pair containing the sample G1 (Table 3.7). The discs after finish milling M1 and M2 again led to the smallest resistance to motion (Fig. 3.6). Similar to the tests D, wear of the balls was much higher than wear of the discs and inversely correlated with it. The highest wear value of the frictional pair was observed for the disc sample VB2.

Table 3.8 presents the result of the test F. For nine analysed frictional pairs, the smallest values of the friction coefficient were observed for the discs after finish milling—Fig. 3.7; in these cases, small wear values were also observed. Coefficients

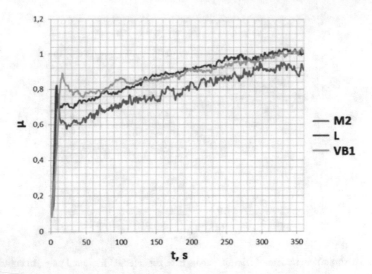

Fig. 3.9 Maximum coefficient of friction versus time for selected frictional pairs, in conditions H

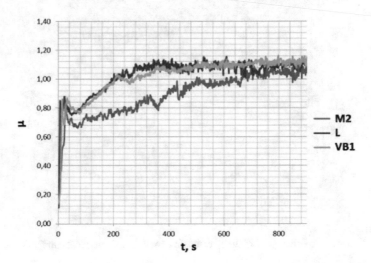

Fig. 3.10 Maximum coefficient of friction versus time for selected frictional pairs, in conditions I

of friction corresponding to samples after grinding G1 were also small, and wear of balls was much higher than wear of discs except for the sample VB2. Coefficients of friction were proportional to total wear (linear coefficients of correlation were about 0.7).

Similar to the tests A–F the lowest coefficient of friction in the test G was found for finish-milled samples (Table 3.9). The coefficient of friction became stable after approximately 450 s (Fig. 3.8). Wear of balls was much higher than wear of discs.

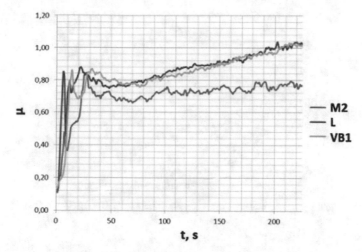

Fig. 3.11 Maximum coefficient of friction versus time for selected frictional pairs, in conditions J

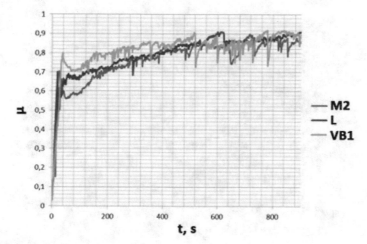

Fig. 3.12 Maximum coefficient of friction versus time for selected frictional pairs, in conditions K

Total wear of frictional pairs containing lapped and polished discs was higher in comparison with vapour blasted samples. Similar tendency was also observed under the F conditions.

It was noticed after the test H (Table 3.10) that the smallest values of wear and coefficient of friction were obtained for disc samples after finish milling process M1 and M2. Friction coefficients corresponding to lapped and polished discs were higher than obtained for vapour blasted surfaces. The coefficient of friction increased during the test (Fig. 3.9). A proportional dependence between wear and the coefficient

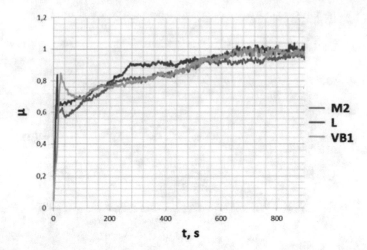

Fig. 3.13 Maximum coefficient of friction versus time for selected frictional pairs, in conditions L1

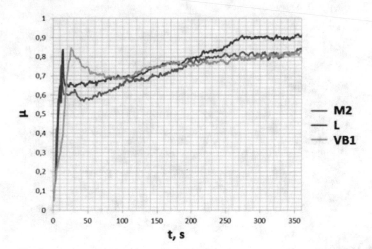

Fig. 3.14 Maximum coefficient of friction versus time for selected frictional pairs, in conditions M

of friction was observed; the linear coefficients of correlation were in the range: 0.75–0.8. Wear of balls was higher and inversely proportional to wear of discs.

When the frequency was 80 Hz and the normal load was 45 N (test I), the smallest values of the friction coefficient were found for finish-milled discs, especially M2 (Table 3.11). The coefficient of friction typically became stable after 400 s, but for frictional pairs having disc surfaces after finish milling, such stabilisation was reached later (Fig. 3.10). Wear of balls was higher than wear of discs and in most cases (except for the samples with the largest surface height M3 and VB2) was

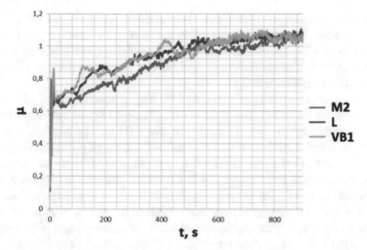

Fig. 3.15 Maximum coefficient of friction versus time for selected frictional pairs, in conditions N

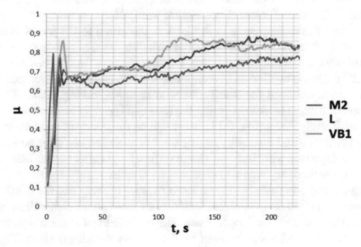

Fig. 3.16 Maximum coefficient of friction versus time for selected frictional pairs, in conditions O

negative (build-ups and/or material transfer). The highest wear was observed for the frictional pair containing polished disc surface, while the smallest for the vapour blasted sample VB1. Wear of balls was inversely proportional to wear of discs.

After the test J (Table 3.12), the smallest values of the coefficient of friction were observed for the disc samples after finish milling M1 and M2. Low value of the average coefficient of friction was also obtained for the frictional pair containing rough-milled disc M3. The coefficient of friction increased as the test progressed (Fig. 3.11). The wear volume of frictional pairs containing milled discs surfaces was low. Wear of balls was higher than wear of discs. Again, higher values of wear were obtained for frictional pairs containing disc surfaces after lapping and polishing compared to samples after vapour blasting. Wear of discs was inversely correlated with wear of balls. The coefficient of friction was proportional to wear (the linear coefficient of correlation was 0.6).

For the smallest frequency of 20 Hz and the largest normal load of 75 N (the test K), wear of ball was higher than wear of disc (Table 3.13) and inversely correlated with it (the linear coefficient of correlation was −0.61). The smallest wear value was obtained for disc surfaces after finish milling and rough grinding G1. However, no negative wear of balls was observed. The smallest coefficients of friction were obtained for frictional pairs containing finish-milled and ground disc surfaces. The average coefficients of friction corresponding to the discs with large surface height M3, VB1 and VB2 were higher than those obtained for sliding pairs with smoother discs. This is due to higher frictional resistance of rubbing pairs with rough disc surfaces in the initial stage of the test (Fig. 3.12). The coefficient of friction typically obtained stable value after 500 s. It seems that an increase in the load caused the reduction in the effects of the disc surface topography on the frictional resistance. The correlation between the final coefficient of friction and total wear was also observed (the linear coefficient of correlation was 0.83).

An increase in the frequency in the L1 test compared to the test K resulted in increases in wear and coefficient of friction. Similar to cases presented previously, wear of ball was higher than wear of disc. The coefficient of friction was proportional to wear, and the linear coefficient of correlation was about 0.8. The lowest coefficients of friction were obtained for disc surfaces after finish milling (Table 3.14). The coefficients of friction increased during the tests and typically reached stable values after approximately 700 s (Fig. 3.13). At the same frequency of oscillation, higher frictional resistance and especially higher wear of smooth discs after polishing and lapping in comparison with discs with large surface height after vapour blasting were noticed. Similar tendency was also observed for smaller normal loads.

During the test M (Table 3.15), the lowest values of wear and the coefficient of friction were obtained for assemblies with disc surfaces after finish milling whereas the largest for polished and lapped ones. The coefficient of friction was inversely proportional to wear. The coefficient of friction increased during the tests (Fig. 3.14). It was found that wear of balls was higher and inversely proportional to wear of discs. No negative values of discs wear were observed.

Wear of balls in the test N (Table 3.16) was higher than wear of discs. The smallest average coefficients of friction were obtained for finish-milled surfaces (especially M2), while small values of the final coefficient of friction were obtained for discs M1 and M2 and for discs after vapour blasting, especially VB2. The coefficient of friction increased during the test and probably reached a stable value in the final part of the test (after about 800 s—Fig. 3.15). The relationship previously observed at higher frequency of oscillations was confirmed—the low wear and coefficients of friction were obtained for rough disc surfaces after vapour blasting in comparison with smooth samples after polishing and lapping. The coefficient of friction was proportional to the total wear (the linear coefficient of correlation was about 0.7).

Similar to previous studies, during the tests O wear of balls was higher than wear of discs (Table 3.17). No negative values of disc wear were observed. The smallest values of the friction coefficient were obtained for the rubbing pair with the sample M2. Small values of the coefficient of friction were also obtained for frictional pairs with discs M1, M3 and G1. The coefficient of friction increased as the tests progressed (Fig. 3.16). Wear of discs was inversely proportional to wear of balls. The coefficient of friction was inversely proportional to wear. The total wear volumes of all frictional pairs were similar.

On the basis of presented research, it was observed that the smallest values of the coefficient of friction were obtained for frictional pairs containing disc surfaces after finish milling, when the ball's movement was perpendicular to the main texture direction (lay). Similar results were obtained during earlier research, reported in [1]; in this work, it was also found that when the movement of the ball was parallel to the disc lay, the resistance to motion was much higher, compared to perpendicular position. Small frictional resistance was also often obtained for frictional pairs containing discs after rough grinding and sometimes after rough milling. Wear of balls was higher than wear of discs, which was caused by accumulation of oxidised wear products at the bottom of wear scars on the disc surface, which protected the disc from wear and increased wear of balls. This behaviour was confirmed during the analysis of chemical constitution of wear scars of the disc of 47 HRC hardness after dry gross fretting [5]. Wear of balls was inversely proportional to wear of discs. In many cases, the wear volume of the tribological system was proportional to the frictional resistance. It is interesting to compare the properties of frictional pairs containing smooth discs after lapping and polishing, and the discs with large surface height. For the smallest normal load of 15 N, wear of smooth discs and total wear of frictional pairs with these discs were smaller regardless of the frequency. For the average load of 45 N, tendency to lower frictional resistance and wear for frictional pairs containing discs after vapour blasting at higher frequency of oscillation 50 and 80 Hz was observed. This tendency was also found for the highest normal load, except for the test O; however, at the smallest frequency, smooth discs corresponded to lower initial and average values of the coefficient of friction. It was also observed that when the normal load increased the effect of surface topography on the coefficient of friction decreased.

Figures 3.17 and 3.18 show contour plots of selected disc and ball samples after tests. Wear had abrasive and adhesive mechanisms. For the rough discs VB1 and VB2, the plastic deformations were also observed. One can see an increase in wear of the tribological system at increasing frequency. This is due to an increase in the number of cycles. For the disc surfaces VB1 and VB2, the largest areas of the wear scars with the lowest normal load were observed.

Figures 3.19 and 3.20 show examples of fretting loops together with calculated values of the slip index δ. Based on them, it can be concluded that during the tests a gross fretting occurred. An increase in the normal load caused a decrease in the slip index.

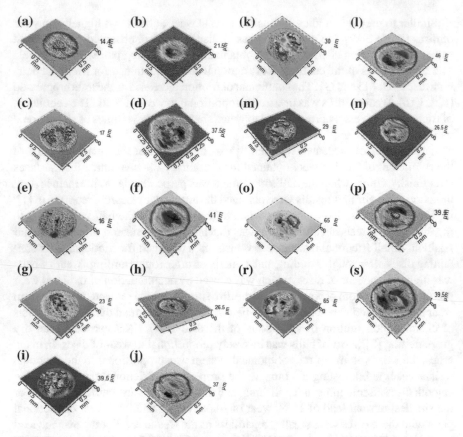

Fig. 3.17 Isometric views of scars of the disc L (**a, c, e, g, i, k, m, o, r**) and ball (**b, d, f, h, j, l, m, p, s**) after the tests A (**a, b**), B (**c, d**), D (**e, f**), F (**g, h**), G (**i, j**), I (**k, l**), K (**m, n**), L1 (**o, p**), N (**r, s**)

During analysis of the effect of the frequency and the normal force on the coefficient of friction and wear, two variants were studied:

– the first, with constant test duration: conditions A, B, D, F, G, I, K, L1, N
– the second, with constant number of cycles: A, C, E, F, H, J, K, M, O.

At the smallest normal load, the changes of the friction coefficient in time depend on frequency. At the lowest frequency, the coefficient of friction became stable after about 400 s; however, at higher frequencies, it reached a maximum after about 200 s, and then, it decreased. For the middle normal load in the first variant, the course of the friction coefficient in time did not depend significantly on the frequency: the coefficient of friction increased and reached a steady-state value (sometimes for the discs after finish milling the coefficient of friction did not reach a stable value) but in the second variant at higher frequencies the coefficient of friction increased. However, when the normal load was the highest, the course of the friction coefficient in time also did not depend on the frequency: the coefficient of friction after abrupt initial

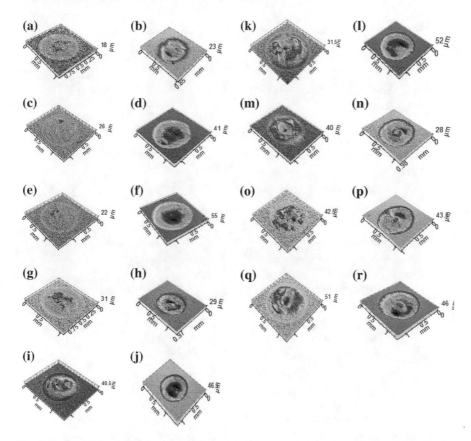

Fig. 3.18 Isometric views of scars of the disc VB1 (**a, c, e, g, i, k, m, o, r**) and ball (**b, d, f, h, j, l, m, p, s**) after the tests A (**a, b**), B (**c, d**), D (**e, f**), F (**g, h**), G (**i, j**), I (**k, l**), K (**m, n**), L1 (**o, p**), N (**r, s**)

changes slowly increased with possible stabilisation after about 800 s. However, at constant frequency, an increase in the normal load typically caused growth of time necessary to obtain a stable value of the coefficient of friction.

In the first variant, an increase in the frequency caused an increase in both the coefficient of friction and wear, regardless of the normal load applied. On the other hand, an increase in the normal load caused a reduction in the coefficient of friction regardless of the frequency of oscillations. An increase in the normal load also caused an increase in wear at the smallest frequency of oscillations. At a frequency of 50 Hz, increasing the normal load from 15 to 45 N caused an increase in wear; however, a further increase in the normal load did not cause any changes in wear. At the highest frequency of oscillation, the change of the normal load had no significant effect on wear of the elements of the analysed tribological system. These results are interesting. They suggest that at higher speed, load changes do not have significant effects on wear. It was also found that wear of disc was positive only at the lowest

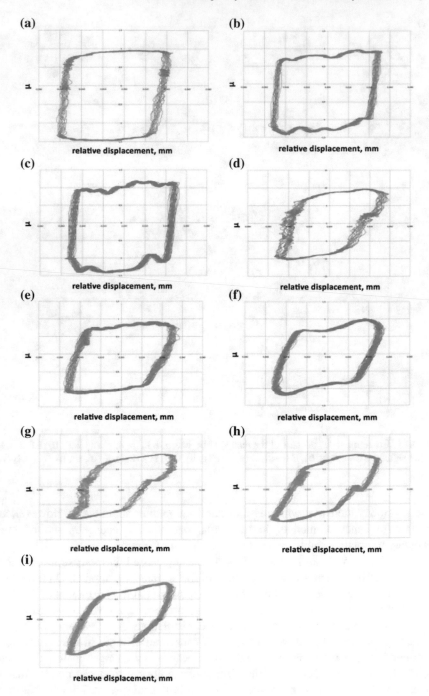

Fig. 3.19 Fretting loops for frictional pairs containing the disc L after the tests $A - \delta = 9.35$ (**a**), $B - \delta = 9.8$ (**b**), $D-\delta = 9.69$ (**c**), $F-\delta = 3.58$ (**d**), $G-\delta = 4.30$ (**e**), $I-\delta = 4.19$ (**f**), $K-\delta = 2.26$ (**g**), $L1-\delta = 2.33$ (**h**), $N - \delta = 2.33$ (**i**)

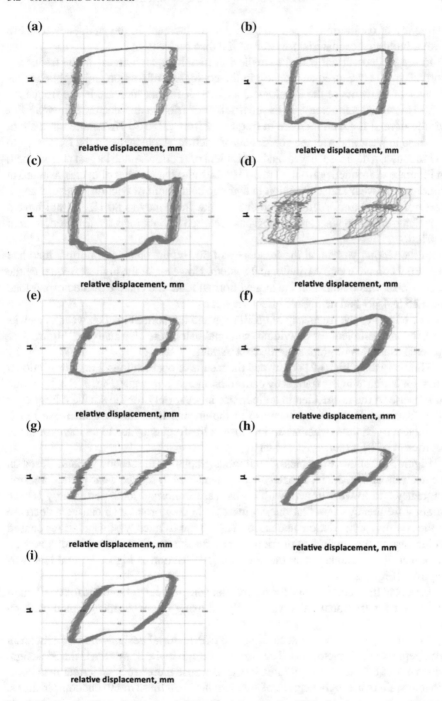

Fig. 3.20 Fretting loops for frictional pairs containing the disc VB1 after the tests A − δ = 8.39 (**a**), B − δ = 7.94 (**b**), D − δ = 9.75 (**c**), F − δ = 3.3 (**d**), G − δ = 4.88 (**e**), I − δ = 4.21 (**f**), K − δ = 2.67 (**g**), L1 − δ = 1.98 (**h**), N − δ = 2.03 (**i**)

frequency of oscillation. It can be seen that an increase in the frequency tends to form a build-ups or material transfer on the disc surfaces.

In the second variant with the smallest normal load, an increase in the frequency from 20 to 50 Hz caused an increase in the coefficient of friction and slight changes in wear of elements of the tribological system. A further increase in the frequency to 80 Hz caused an increase in the coefficient of friction and decrease in wear. For middle normal load, an increase in frequency from 20 to 50 Hz resulted in a slight decrease in wear and average coefficient of friction. An increase in the frequency to 80 Hz caused a decrease in wear and coefficient of friction. At the largest normal load, an increase in frequency from 20 to 50 Hz for the same number of cycles resulted in a decrease in wear and a reduction in average coefficient of friction. These changes were relatively large. A further increase in the frequency to 80 Hz did not have a significant effect on wear and in most cases caused a slight increase in the coefficient of friction.

In the second variant, at the frequency of 50 Hz, the change of normal load had little effect on wear of the tribological system. However, at the highest frequency the largest wear was obtained at the largest normal load; the change of the normal load from 15 to 45 N had negligible effect on wear.

The effects of the frequency of oscillation and the normal load on wear as well as the friction coefficient for individual frictional pairs were also analysed. Figure 3.21 presents examples of graphs for the first variant.

The analysis of Fig. 3.21 confirmed the results of observations. For both frictional pairs, an increase in the frequency caused an increase in wear (Fig. 3.21a, b), while an increase in the normal led to an increase in wear only for the smallest frequency of oscillations. However, an increase in the normal load caused a decrease in the coefficient of friction, whereas an increase in the frequency led to an increase in the coefficient of friction (Fig. 3.21c–f).

Figure 3.22 shows examples of analogous graphs for the second variant. Based on the analysis of Fig. 3.22a, b, one can see a tendency to reduce wear with increasing frequency and to wear growth with an increase in normal load, particularly for the smallest frequency of oscillation. An increase in the normal load caused a decrease in the coefficient of friction, and an increase in the frequency of oscillations caused an increase in the coefficient of friction only for the smallest normal load; however, the effect of the frequency on the coefficient of friction at higher normal load was opposite (Fig. 3.22c–f).

Effects of the frequency and the normal load on the wear volume and the coefficient of friction for frictional pairs were analysed using the experimental methodology [6, 7].

Research was carried out according to the 3^k factorial design—that is, a factorial arrangement with k factors [7]. The simplest design in the 3^k system is the 3^2 design, which has two factors, each at three levels. Because there are $3^2 = 9$ treatment combinations, there are eight degrees of freedom between these treatment combinations.

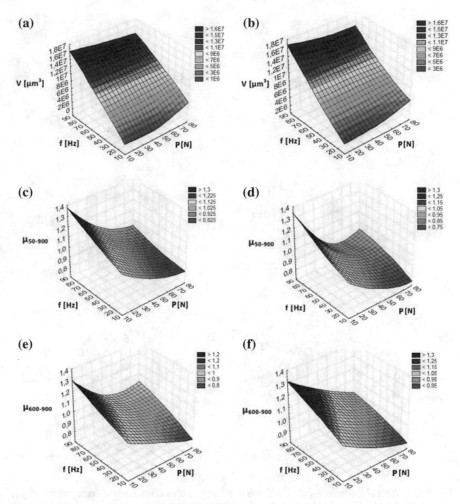

Fig. 3.21 Effect of the normal load P and the frequency of oscillation f on the wear volume V_{total}, the average coefficient of friction μ_{50-900}, the final coefficient of friction $\mu_{600-900}$, for frictional pairs containing discs M2 (**a**) and VBI2 (**b**) for the first variant (constant test duration)

This plan allows to obtain the mathematical model of the studied process in the form of a polynomial of the second degree

$$y = b_0 + \sum b_k x_k + \sum b_{kk} x_k^2 + \sum b_{kj} x_k x_j$$

where

$x_1, \ldots, x_k$—input factors (variable parameters)
$b_0, \ldots, b_k$—coefficients of a polynomial regression
y—regression function (measured value)

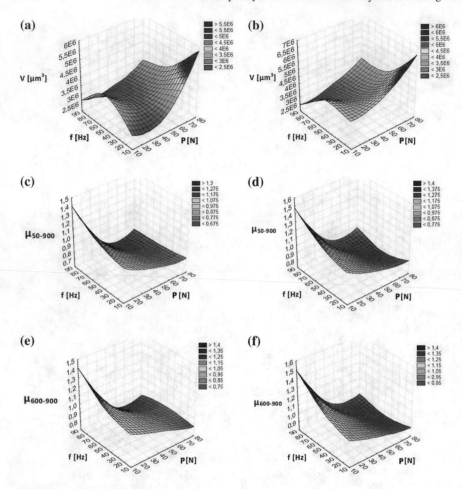

Fig. 3.22 Effect of the normal load P and the frequency of oscillation f on the wear volume V_{total}, the average coefficient of friction μ_{50-900}, the final coefficient of friction $\mu_{600-900}$, for frictional pairs containing discs M2 (**a**) and VBI2 (**b**) for the second variant (constant number of cycles)

In three-level plans, the input factors assume values on three levels of variation [6]:

high (marked as +1)
intermediate (marked as 0)
low (marked as −1)

Based on the three-level plan, an analysis of the repeatability of experiments and the adequacy of the obtained regression equations was carried out. Calculations were made for each of the tested input factors, thus obtaining dependencies, showing the effect of selected parameters of the fretting process on the coefficient of friction and volumetric wear of the frictional pair.

For tests of frictional pairs containing the disc M2, the following regression equations were obtained (the average coefficient of friction was denoted by μ_{av}, and the final one by μ_f:
variant I:

$$V_{total} = -2277710.1 + 263478.1 f - 548.8 f^2$$
$$+ 22557.3 P - 8883.1 f P + 242.9 P^2$$

$$\mu_{av} = 1.1 + 0.004 f - 0.00001 f^2 - 0.01 P$$
$$- 0.00002 f P + 0.0001 P^2$$

$$\mu_f = 0.8 + 0.001 f + 0.00001 f^2$$
$$- 0.001 P + 0.00006 P^2$$

variant II:

$$V_{total} = 3263304.6 + 22291.8 f - 146.7 f^2$$
$$- 37252.6 P - 396.4 f P + 765.008 P^2$$

$$\mu_{av} = 1.2 + 0.002 f + 0.00001 f^2 - 0.02 P$$
$$- 0.00007 f P + 0.0001 P^2$$

$$\mu_f = 1.2 + 0.003 f - 0.01 P$$
$$- 0.00006 f P + 0.0001 P^2$$

The following regression equations were obtained for the frictional pairs containing the VB2 disc:
variant I:

$$V_{total} = 733903.2 + 144422.6 f + 434.7 f^2$$
$$+ 19700.3 P - 774 f P + 211.1 P^2$$

$$\mu_{av} = 1.2 + 0.003 f - 0.01 P$$
$$- 0.00003 f P + 0.00006 P^2$$

$$\mu_f = 1.1 + 0.004 f - 0.000006 f^2$$
$$- 0.005 P - 0.00001 f P + 0.00002 P^2$$

variant II:

$$V_{\text{total}} = 5043008.7 + 30199\,f + 144\,f^2$$
$$- 25941\,P - 182.3\,f\,P + 523.8\,P^2$$

$$\mu_{\text{av}} = 1.3 - 0.0002\,f + 0.00004\,f^2$$
$$- 0.01\,P - 0.00006\,f\,P + 0.0001\,P^2$$

$$\mu_{\text{f}} = 1.2 + 0.001\,f + 0.00003\,f^2$$
$$- 0.01\,P - 0.00007\,f\,P + 0.00009\,P^2$$

After the analysis of the regression equations, one can find their adequacy to the test results. In the first variant, wear was proportional to the frequency, while coefficients of friction were proportional to the frequency and inversely proportional to the normal load. However, in the second variant, larger normal load corresponded to smaller frictional resistance.

In the technical literature, there are available results of research on the effect of the frequency of oscillation on wear in the gross fretting conditions. In these works, it was found that an increase in the frequency had little effect on wear [8] or led to wear reduction [9–12]. These findings are consistent with the results presented above. However, they are contrary to the results obtained by Park et al. [13, 14]. There is small amount of information on the effect of the displacement frequency on the change of the coefficient of friction. The authors of papers [10, 15] found that an increase in the frequency caused an increase in the coefficient of friction. These findings are consistent with the results of the current tests at the constant test duration and at the constant number of cycles for the smallest normal load.

References

1. A. Lenart, P. Pawlus, A. Dzierwa et al., The effect of surface topography on dry fretting in the gross slip regime. Arch. Civ. Mech. Eng. **17**(4), 894–904 (2017). https://doi.org/10.1016/j.acme.2017.03.008
2. M. Varenberg, I. Etsion, E. Altus, Theoretical substantiation of the slip index approach to fretting. Tribol. Lett. **19**, 263–264 (2005). https://doi.org/10.1007/s11249-005-7442-8
3. M. Varenberg, I. Etsion, G. Halperin, Slip index: a new unified approach to fretting. Tribol. Lett. **17**(3), 569–573 (2003). https://doi.org/10.1023/B:TRIL.0000044506.98760.f9
4. P. Pawlus, Simulation of stratified surface topographies. Wear **264**, 457–463 (2008). https://doi.org/10.1016/j.wear.2006.08.048
5. A. Lenart, P. Pawlus, A. Dzierwa et al., The effect of surface texture of steel disc on friction and fretting wear. Tribologia **4**, 39–48 (2018). https://doi.org/10.5604/01.3001.0012.7527
6. C. Koukouvinos, Orthogonal 2 and 3k factorial designs constructed using sequences with zero autocorrelation. Stat. Probab. Lett. **28**, 59–63 (1996). https://doi.org/10.1016/0167-7152(95),00082-8
7. D.C. Montgomery, *Design and Analysis of Experiments*, 8th edn. (Wiley, Hoboken, 2017)

8. S. Soderberg, U. Bryggman, T. McCullough, Frequency effects in fretting wear. Wear **110**, 19–34 (1986). https://doi.org/10.1016/0043-1648(86),90149-3
9. W. Huang, B. Hou et al., Fretting wear behavior of AZ91D and AM60B magnesium alloy. Wear **260**, 1173–1178 (2006). https://doi.org/10.1016/j.wear.2005.07.023
10. E.R. Leheup, D. Zhang, J.R. Moon, Fretting wear of sintered iron under low normal pressure. Wear **221**, 86–92 (1998). https://doi.org/10.1016/S0043-1648(98),00265-8
11. L. Li, I. Etsion, F.E. Talke, The effect of frequency on fretting in a micro-spherical contact. Wear **270**, 857–865 (2011). https://doi.org/10.1016/j.wear.2011.02.014
12. L. Toth, The investigation of the steady stage of steel fretting. Wear **20**, 277–286 (1972). https://doi.org/10.1016/0043-1648(72),90409-7
13. Y.W. Park, G.R. Bapu, K.Y. Lee, Studies of tin coated brass contacts in fretting conditions under different loads and frequencies. Surf. Coat. Technol. **201**, 7939–7951 (2007). https://doi.org/10.1016/j.surfcoat.2007.03.039
14. Y.W. Park, T.S. Narayanan, K.Y. Lee, Effect of fretting amplitude and frequency on the fretting corrosion behaviour of thin plated contacts. Surf. Coat Technol. **201**, 2181–2192 (2006). https://doi.org/10.1016/j.surfcoat.2006.03.031
15. J.D. Lemm, A.R. Warmuth, S.R. Pearson et al., The influence of surface hardness on the fretting wear of steel pairs—its role in debris retention in contact. Tribol. Int. **81**, 258–266 (2015). https://doi.org/10.1016/j.triboint.2014.09.003

Concluding Remarks

Effects of operating parameters (frequency and normal load) and of surface texture of steel disc samples in contact with steel harder ball on wear and the coefficient of friction under dry gross fretting conditions were studied. During analysis of the effect of the frequency and the normal force on the coefficient of friction and wear, two variants were taken into consideration: with constant test duration and with the constant number of cycles. Nine types of one-process disc surfaces were studied.

It was found that an increase in the normal load caused a reduction in the coefficient of friction and an increase in wear mainly at the lowest frequency of oscillation.

When the test duration was constant, an increase in the frequency caused an increase in the frictional resistance and wear. An increase in the frequency above 20 Hz resulted in a tendency to form build-ups or material transfers on the tested surfaces.

For the constant number of cycles, an increase in the frequency of oscillations mainly caused a reduction in wear. However, the effect of the frequency on the coefficient of friction depended on the normal load. At the lowest load, an increase in the frequency caused an increase in the coefficient of friction. At higher normal loads, an increase in the frequency primarily led to a reduction in the coefficient of friction.

The lowest values of the coefficient of friction in the whole test range were obtained for frictional pairs containing finish-milled discs when the direction of ball's movement was perpendicular to the main disc texture direction. Small frictional resistance was often obtained for the frictional pair containing the disc after rough grinding and sometimes after rough milling. Wear of balls was larger and inversely proportional to wear of discs. In many cases, the wear volume of the tribological system was proportional to the frictional resistance. At higher oscillation frequencies of 50 and 80 Hz and the normal loads above 15 N, there was a tendency to lower frictional resistance and wear for frictional pairs containing discs after vapour blasting with large surface height compared to frictional pairs containing smooth discs after polishing and lapping. With an increase in the normal load, the effect of the disc surface topography on the coefficient of friction decreased.

P. Pawlus et al., *Dry Gross Fretting of Rough Surfaces*, Manufacturing and Surface Engineering, https://doi.org/10.1007/978-3-030-31563-4

Printed in the United States
By Bookmasters